CAMBRIDGE LIBRARY COLLECTION

Books of enduring scholarly value

Botany and Horticulture

Until the nineteenth century, the investigation of natural phenomena, plants and animals was considered either the preserve of elite scholars or a pastime for the leisured upper classes. As increasing academic rigour and systematisation was brought to the study of 'natural history', its subdisciplines were adopted into university curricula, and learned societies (such as the Royal Horticultural Society, founded in 1804) were established to support research in these areas. A related development was strong enthusiasm for exotic garden plants, which resulted in plant collecting expeditions to every corner of the globe, sometimes with tragic consequences. This series includes accounts of some of those expeditions, detailed reference works on the flora of different regions, and practical advice for amateur and professional gardeners.

Conversations on Vegetable Physiology

Jane Haldimand Marcet (1769–1858) wrote across a range of topics, from natural philosophy to political economy. Her educational books were especially intended for female students, to combat the prevalent idea that science and economics were unsuitable for women, but they found broader popularity: Michael Faraday, as a young bookbinder's apprentice, credited Marcet with introducing him to electrochemistry. This two-volume work, first published in 1829, is a beginner's guide to botany. Since the chief aim was accessibility, Marcet does not dwell on the often burdensome process of plant classification, but focuses on plant forms and botany's practical applications. She presents the facts in the form of simple conversations between two students and their teacher. Based on the lectures of the Swiss botanist Candolle, Volume 2 considers agriculture and plant diseases, the cultivation of trees and culinary vegetables, and the effects of humans on flora.

Cambridge University Press has long been a pioneer in the reissuing of out-of-print titles from its own backlist, producing digital reprints of books that are still sought after by scholars and students but could not be reprinted economically using traditional technology. The Cambridge Library Collection extends this activity to a wider range of books which are still of importance to researchers and professionals, either for the source material they contain, or as landmarks in the history of their academic discipline.

Drawing from the world-renowned collections in the Cambridge University Library and other partner libraries, and guided by the advice of experts in each subject area, Cambridge University Press is using state-of-the-art scanning machines in its own Printing House to capture the content of each book selected for inclusion. The files are processed to give a consistently clear, crisp image, and the books finished to the high quality standard for which the Press is recognised around the world. The latest print-on-demand technology ensures that the books will remain available indefinitely, and that orders for single or multiple copies can quickly be supplied.

The Cambridge Library Collection brings back to life books of enduring scholarly value (including out-of-copyright works originally issued by other publishers) across a wide range of disciplines in the humanities and social sciences and in science and technology.

Conversations on Vegetable Physiology

*Comprehending the Elements of Botany,
with their Application to Agriculture*

VOLUME 2

JANE HALDIMAND MARCET

CAMBRIDGE
UNIVERSITY PRESS

University Printing House, Cambridge, CB2 8BS, United Kingdom

Published in the United States of America by Cambridge University Press, New York

Cambridge University Press is part of the University of Cambridge.
It furthers the University's mission by disseminating knowledge in the pursuit of
education, learning and research at the highest international levels of excellence.

www.cambridge.org
Information on this title: www.cambridge.org/9781108067461

© in this compilation Cambridge University Press 2014

This edition first published 1829
This digitally printed version 2014

ISBN 978-1-108-06746-1 Paperback

This book reproduces the text of the original edition. The content and language reflect
the beliefs, practices and terminology of their time, and have not been updated.

Cambridge University Press wishes to make clear that the book, unless originally published
by Cambridge, is not being republished by, in association or collaboration with, or
with the endorsement or approval of, the original publisher or its successors in title.

CONVERSATIONS

ON

VEGETABLE PHYSIOLOGY.

VOL. II.

CONVERSATIONS

ON

VEGETABLE PHYSIOLOGY ;

COMPREHENDING

THE ELEMENTS OF BOTANY,

WITH

THEIR APPLICATION TO AGRICULTURE.

BY THE AUTHOR OF

" CONVERSATIONS ON CHEMISTRY," " NATURAL PHILOSOPHY'
&c. &c.

IN TWO VOLUMES.

VOL. II.

LONDON :

PRINTED FOR
LONGMAN, REES, ORME, BROWN, AND GREEN,
PATERNOSTER-ROW.
1829.

LONDON:
Printed by A. & R. Spottiswoode,
New Street-Square.

TABLE OF CONTENTS.

VOL. II.

A 3

CONVERSATION XIX.

ON FRUIT.

CONVERSATION XX.

ON THE SEED.

CONVERSATION XXV.

ON THE INFLUENCE OF CULTURE ON VEGETATION.

CONVERSATION XXVI.

ON THE DEGENERATION AND THE DISEASES OF PLANTS.

CONVERSATIONS.

ON THE MULTIPLICATION OF PLANTS.

CONVERSATION XV.

ON THE PROPAGATION OF PLANTS BY SUBDIVISION.

MRS. B.

I T is now time to turn our attention from the preparation of the soil to the study of the plants which are to be raised in it.

CAROLINE.

After having provided suitable accommodation for their reception, and an abundant store of food for their subsistence, they will no doubt increase and multiply with rapidity.

MRS. B.

That is not all. If we have taken so much pains to provide for the welfare of the vegetable creation, it is with the interested view of its affording us food and raiment; we shall therefore select for cultivation such plants as are best suited to that purpose.

There are two modes of propagating vegetables: the first consists in subdividing the parts of a plant, so that from one individual several may be formed; the second mode is that of raising new plants by the germination of the seed.

In order to be able in every case to distinguish these two processes, you must observe that the seed is always contained in an envelope, and that it is prepared by organs exclusively destined for that purpose. These organs compose the flower or blossom. Now the plant which results from the germination of the seed, is always of the same species as that in which the seed originated; but varying from it frequently in the quality of its fruit, and not inheriting any of the peculiarities which may have casually distinguished the individual parent-plant.

When, on the contrary, a new plant is raised by separating from the parent-stock a slip or a layer, you not only produce an individual of the same species, but, if I may so express it, a continuation of the same plant, possessing every peculiarity by which it may casually be distinguished from others of its species.

EMILY.

When these peculiarities are of an advantageous nature, it must be desirable to raise the plant by division; otherwise, I suppose, it is more easily accomplished by sowing the seed.

MRS. B.

But the process is much more tardy. A seedling tree of ten years' growth will perhaps not be more advanced than one raised by a slip of five years old; then, when you are provided with a plant which bears remarkably fine fruit, you are sure that if propagated by division it will produce fruit of equally good quality. This mode affords, therefore, the most certain means of improving the species.

CAROLINE.

Reproduction by seed is the mode adopted by Nature; that by division the invention of art.

MRS. B.

The latter is also sometimes employed by Nature, as you will see.

Reproduction by division tends to diminish the quantity of seed. The vine, which in a state of nature, bears five seeds in each grape, when propagated by this mode, has only two; and some vines lose them entirely, so as to leave no possibility of reproducing the plant but by division.

EMILY.

The fruit no doubt profits by this deficiency of seed, as the sustenance which would go to ripen the seed, will be expended in enriching the juices of the grape. I have observed that apples and oranges, which have the fewest pips, are the highest flavoured.

MRS. B.

The remark is applicable to fruits in general. The sugar-cane, propagated by division, wholly loses its seed; and so do also the succulent plants of the Cape of Good Hope, after having been for a number of years transplanted into Europe.

CAROLINE.

But I cannot comprehend how a slip can strike root. That root, branch, and every part of a plant should be developed by the germination of a seed, in which it existed in a latent state, is easy to conceive; but that a root should grow from the extremity of a young shoot seems to favour the idea of casual reproduction, which is not to be met with in Nature.

MRS. B.

There is reason to suppose that germs, in some respect analogous to those which are contained in the seed, exist in almost every part of a plant, but are not developed unless placed under favourable circumstances; that these germs are of two distinct species, the one producing stems, the other roots. The former originate chiefly in the axilla of the leaf, or part which unites the leaf to the stem ; and which, from the analogy it bears to the union of the arm to the body in the human frame, is called the axilla : the latter shoot out roots on each side.

It has been affirmed that roots may, by exposure to the air, be converted into branches; and branches, by being buried in the earth, transformed into roots ; and this, as I believe I before mentioned to you, has been attempted to be proved by overturning a willow, burying the head in the ground, and leaving the roots upwards exposed to the air. But what was the result? Not that the branches became roots, and the roots branches ; for the former being unfurnished with the organs of absorption, and the latter with those of evaporation, it was impossible for them to exchange their respective functions; but the branches being deprived under ground both of light and air, and of all the circumstances favourable to the developement of the germs of other branches, these branches do not shoot. The same circumstances being, on the other hand, particularly adapted to the developement of the germs of roots, these strike out into the soil.

In the mean time the roots, which have been compelled to change places with the branches, being exposed to the light and air, and so situated as to favour the developement of the germs of branches, and in direct opposition to that of the germs of roots, shoot out young branches from their naked roots, and in the course of time cover them with foliage.

I recollect having seen the leaf of a plant, which, when simply laid upon moist ground, struck out roots from its edges into the soil.

This is the Bryophyllam: the flower is the only part of a plant which is incapable of developing either a root or a stem, except through the medium of the seed, the production of which is its sole and exclusive function.

There are three modes of multiplying plants by division : —

The first by layers ;

The second by scions, or slips ;

The third by grafts.

When you intend to multiply by subdivision, you place that portion of the plant which you intend to separate from the remainder under such circumstances as are requisite to enable it to develope the organs in which it is deficient, and which are necessary to its independent existence. If it be a branch, the organ wanting is a root; if it be a root, the organ necessary to be developed is a stem.   How is this to be accomplished ?

You must, I suppose, bury the extremity of the

branch in moist ground, to favour the developement of roots; and, in the other case, train the roots above ground to encourage that of branches.

MRS. B.

Exactly. It is the cambium, you must recollect, which, in its retrograde course through the liber, and partly through the alburnum, nourishes these germs; if, therefore, you propose to develope them in any particular part of the plant, you must accumulate the cambium in that spot. This may be done in several different ways. In the first place, you may make an annular incision in the bark or rind, and, by thus impeding the descent of the cambium, accumulate it in the upper section, where it will produce a swelling or protuberance of the bark. The germs situated in the neighbourhood of this rich magazine of food, if in other respects favourably circumstanced, are developed; that is to say, if the annular incision be exposed to light and air, the germs of branches will shoot; if below ground, those of roots will strike into the soil. Indeed any casual interference with the descent of the cambium is almost immediately followed by the sprouting of a bud. In order to make a layer, you bend down a pliant branch without separating it from the plant, and fasten it in the ground; sometimes a slight incision is made at the spot in which it is confined. — Now, what follows? The cambium, descending through the branch, finds some difficulty in re-

turning to the stem : this obstacle is sufficient to
occasion a small accumulation, and the shooting
out of several germs of roots.

There are some creeping plants which propagate
themselves in this manner without the aid of man.
Their lower branches, trailing upon the ground,
are often partially covered with earth washed over
them by the rain : if, in this operation, they are
slightly wounded by friction, or the contact of any
hard substance, such as gravel or pebbles, the free
passage of the cambium is interrupted, roots strike
out, and the branch which connected them with
the parent-stock, being in a great measure deprived
of its nourishment by the young roots, rots and
perishes ; the separation is thus made, and the requisite organs being developed, the layer becomes
a new individual plant.

CAROLINE.

I have seen carnations and ranunculuses thus
propagated ; and I am delighted to hear the explanation of an operation I have often witnessed
without understanding it.

MRS. B.

Laurels and most evergreens are also propagated
by layers ; and it is the regular mode used in vineyards. A branch of vine is laid under ground, and
the extremity of it raised up above the soil in that
spot where you wish to produce a new plant. If
the branch be long and pliable, several plants may

be made to spring from it. This is called a ser-
pentine layer, because the branch takes a serpen-
tine direction, being made alternately to sink below
and rise above ground, as often as it is intended
that new roots and stems should shoot from it.

Layers are sometimes made in arches by burying the extremity of the branch only; the separation is afterwards made when the branch has
struck root: this mode is particularly suited to
the raspberry and every species of bramble.

CAROLINE.

I have heard that there is a tree in Senegal
called the Mangrove, or Rhizophora, whose
branches, descending to the ground, bury their
extremities in the soil, and strike root, thus forming
beautiful natural arcades around the parent stem.

MRS. B.

Several fig-trees in the East Indies grow and
propagate in the same manner. The ancients
sometimes twisted the branch at the spot where
they wished a root to strike: to this process we
have substituted the more gentle mode of strangulation by ligatures, which injures the branch less,
and yet arrests the cambium sufficiently to produce
an accumulation.

Another mode of making layers consists in slit-
ting the branch from the bottom upwards, and
drawing the portion slit on one side, so as to form
the figure of a Y reversed, the branches being of

unequal length. The portion of the cambium which descends into the slit, finding no vent, accumulates and strikes root.

EMILY.

I have seen the gardener propagate the Magnolia, and other rare and delicate plants, by gently bending some of their most pliant branches to the ground, and covering every part of them with earth excepting their extremities; by this means a considerable number of layers may be obtained at a time.

MRS. B.

Layers are also sometimes made completely above ground, though, it is true, this cannot be done without the aid of the soil; for it is necessary that the branch should be surrounded with moist earth, which may be contained either in a flower-pot or a small basket, having an opening sufficiently large to admit of the branch passing through it.

CAROLINE.

The germs then strike root in this soil. I have seen the Oleander propagated this way.

MRS. B.

M. Humboldt, the celebrated naturalist, when travelling in America, provided himself with strips of coarse pitched cloth, which he substituted in the place of a basket, to confine the earth round branches from which he wished to make layers. He adjusted them round the branches of trees, in forests through which he intended to return some months afterwards, when the roots would have shot out; and by this means he brought over to Europe a number of very curious and valuable new plants, which have not only enriched our botanical gardens, but have been generally disseminated both for use and ornament.

The most favourable season for propagating by layers, in these temperate latitudes, is the latter end of February or the beginning of March. This season is called by gardeners the first spring: it precedes the ascent of the sap, and enables the layers to collect the first drops of cambium which are elaborated. In England and other northern climates, where vegetation is less forward, the end of March, which is called the second spring, is sufficiently early for this purpose. The month of April, in which the budding of the leaf takes place, is denominated the third spring.

The safest way to ensure the success of layers is to leave them a year without separating them from the parent stock, in order to give them the chance of striking root during the ascent of the autumnal sap, if they have failed to do so in the spring.

Both succulent and aqueous plants are very difficult to propagate by layers; because the cambium, instead of forming a protuberance to nourish the germs, runs out and is lost. The operation is more likely to succeed on plants of this description by strangulation than by incision.

The propagation of plants by slips is very analogous to that of layers; indeed the only difference
is, that the branch you destine to become a new
plant is separated from the parent stem previous to
the developement of those organs which are necessary to ensure it an independent existence.

CAROLINE.

I am much better acquainted with this species of
propagation, I have raised so many geraniums by
slips.   Nothing is more easy : you merely cut off
a young branch, and plant it in a pot of earth.
But I am completely ignorant how it lives : whether it absorbs water before it strikes root, or
whether it nourishes the embryo roots by its own
substance.

MRS. B.

I believe no one can boast of having a perfect
knowledge of the process; but I am inclined to
think, that the cambium which descends in the
slip, and which was destined to nourish the lower
part of the branch whence it was cut, finding a sud-
den termination to its course, exudes.   The first
drops fall into the soil; but, from its viscous nature,
those which follow soon coagulate and heal the
wound.   The protuberance then forms, and roots
strike out.

CAROLINE.

This process must, however, be attended with
much greater uncertainty than by layers.   The
slip, being separated from its parent, before it is

able to provide for its own wants, is like a child brought up by hand; whilst the layer is weaned only after it has acquired the power and the means, of finding its own nourishment.

MRS. B.

It is for this reason that the propagation of rare plants is preferable by layers. There are some trees which have such a remarkable facility for sprouting, that whatever part of them you plant in the ground will strike root, be it a branch, the remnant of a stem, or even a simple stake. The willow, the ash, and most trees of white wood, sprout with this facility. The weeping willow is so easily propagated by slips that it is never raised by seed; and all the willows now existing in Europe, and, in all probability, that ever will exist there, are subdivisions of one tree brought originally from Asia.

EMILY.

Greenhouse plants are usually propagated by slips from shoots of the preceding spring ; and sometimes the slip is cut a little below the spring-shoot, so as to include a piece of the shoot of the preceding year.

MRS. B.

This is for the purpose of preventing the extravasation of the cambium, the wood of two years' growth being of a more solid texture.

Branches of three or four years' growth are sometimes planted : they should be placed deep in the soil, to favour the developement of a number of germs. Slips of forked branches are planted with advantage for hedges, as their shoots interlace each other, and form an impenetrable fence.

In raising succulent plants by means of slips, it is necessary either to dry up or cover with mastic the cut end, for the purpose of retaining the cambium : in the isles of France and of Bourbon it is usual to carbonise the ends of slips, in order to prevent its escape.

When you propagate by slips of roots, you must plant them near the surface, in order to facilitate the sprouting of stems. Jessamine, strawberries, and, probably, mushrooms, are propagated in this manner.

EMILY.

I thought that mushrooms were propagated by a species of seed called the spawn.

MRS. B.

The white filaments, vulgarly called the spawn of mushrooms, are in fact the fibres of its roots : these are cut in pieces and sown ; or rather, I should say, planted in a hotbed.

CAROLINE.

In planting potatoes, is it not requisite to leave a spot, called an eye, in each piece? It is from these, I understand, that both stems and branches sprout.

MRS. B.

These eyes are the germs of embryo stems and roots. The potatoe is nothing more than a tubercle formed by an accumulation of cambium in the subterraneous branches of the plant, and destined to nourish the buds which are to be developed the following season. This storehouse of food offers such facility to germination, that when potatoes are heaped in a cellar, of a moderate degree of temperature and moisture, the germs absorb nourishment from the farina of the potatoe and sprout, either roots or stems, according as their situation favours the developement of the one or the other of these germs.

CAROLINE.

In what an enviable situation these germs are placed ! enclosed in a magazine of food — breathing, as it were, an atmosphere of nourishment, and inhaling it at every pore.

EMILY.

Not more so than the germ of a bird, which subsists on the yolk and albumen of the egg until its frame is fully developed.

CAROLINE.

True; it is singular what analogy there is between the different productions of Nature, and

what a fund of knowledge may be derived from
them !

MRS. B.

A fund equally inexhaustible and admirable !
We may consider the works of the creation as a
natural revelation, in which we read the history of
the stupendous operations of the Deity; and which,
the more we study, the more we raise our minds
towards just ideas of their Divine Author, and ele-
vate our hearts by the contemplation of the blessings he has so bountifully lavished upon us.   Not
only are we provided with every thing necessary
to our existence, but care has been taken that even
this, our transitory state, should be rendered agreeable : our food, instead of being insipid or loath-
some, is delightful to the palate; the landscape,
spread before our eyes, instead of being dark or
monotonous, is illumined by a splendid sun, and
variegated by a thousand hues; delicious odours
arise from flowers of enchanting form and colour;
in a word, Nature contains innumerable sources
of enjoyment, which develope and strengthen a
spirit of grateful devotion towards their beneficent
Author.

CONVERSATION XVI.

ON GRAFTING.

MRS. B.

WE may now proceed to the art of grafting, an operation from which we derive our finest fruits. It consists in placing a portion of one plant in juxta-position with another, in such a manner that they shall unite and grow together. The branch which is cut from one tree, to be transferred to another, is called the *graft*, or *scion*, and the tree to which it is transferred the *stock*.

CAROLINE.

This, then, is not a mode of multiplying plants, but of changing their nature; for if a branch of one plant be added to another plant, the number is not increased.

MRS. B.

Certainly not. The advantage of grafting consists in improving the quality, not augmenting the number, of plants. The ancients entertained very exaggerated ideas of this art: they conceived that every species of plant might be grafted on each other; but it is now well ascertained that this operation can be performed only on plants of the same family. To ensure the success of a graft, it is necessary that the vessels of the liber of the two plants should meet and correspond, in order that the cambium should descend from the graft into the stock; for it is by the union of the vessels of the bark of both plants that they are soldered, as it were, together.

CAROLINE.

Then endogenous plants, since they have no bark, cannot be grafted?

MRS. B.

No; at least that operation has not hitherto been performed upon them with success.

Some anatomical analogy is also requisite in the form, the structure, and dimensions of the vessels, which is only to be met with, in plants of the same family. A certain degree of physiological similarity is besides necessary; such, for instance, as that the sap in both plants should rise at the same period. There must also be a correspondence in the size and strength of the plants; for instance, the lilac may be grafted on the ash; but, as the latter has a much greater power of suction, the graft is gorged by the quantity of sap which is thrown up into it, and dies of plethora. If, on the contrary, the ash be grafted on the lilac, the graft perishes for want of nourishment.

A plant which loses its leaves in winter cannot be grafted (at least not without great difficulty) on an evergreen: the latter, absorbing a small quantity of sap during the winter, would send it up into the graft, which would sprout, in a season in which the young shoots would be destroyed by the frost.

CAROLINE.

And if, on the other hand, you were to graft the evergreen on a plant which loses its leaves, the graft would perish of famine.

MRS. B.

Very true; the last analogy required in grafting is, that the two plants should thrive in the same temperature.

When a tree is grafted, the graft will always bear its own fruit, and the tree its own also.

EMILY.

I am surprised at that. Suppose that a branch of cherry were to be grafted on a plum-tree: the sap absorbed by the latter rises through it into the graft, and, being elaborated in the leaves of the branch of the cherry-tree, I should have supposed that it would have changed the nature of the fruit of the plum-tree when in its descent it returns into it.

MRS. B.

No; for though the rising sap is the same for both stock and graft, it is different in its return.

The sap of the stock and that of the graft are each elaborated by their respective leaves, and, when converted into cambium, each supplies nourishment to its own variety.

Mr. Knight has made many ingenious experiments, which tend to show that each variety of fruit requires its own peculiar leaves to bring it to perfection. He grafted several varieties of apples and pears on trees of the same species, and adjusted the grafts close above the flower-buds on the stock: these buds blossomed and bore fruit so long as leaves were suffered to remain on the tree; but in some experiments he stripped them off, so that the sap could be elaborated only in those of the graft, and in those instances the fruit always withered and fell off.

The principal advantage of grafting consists in its affording an easy means of propagating individual plants, which have, either by cultivation or some casual circumstance, attained a high degree of perfection.

EMILY.

This is similar to the advantages obtained by the propagation of plants by layers or slips.

CAROLINE.

I have heard that it is necessary to graft fruit-trees raised from seed, in order to make them bear fruit: yet, if it were so, no fruit would grow wild; and in a state of nature plants could not produce seed to continue their species.

MRS. B.

It is quite erroneous to suppose that seedling fruit-trees will not bear fruit in due time; but this period will be considerably accelerated by grafting. A young tree is not sufficiently strong during the first years of its existence to bear fruit: an apple-tree, for instance, produces none until it has attained the age of ten or twelve years; but, if grafted from a tree that has already borne fruit, it will blossom and produce fruit sometimes as early as the second or third year.

CAROLINE.

Yet grafting cannot increase the age or strength of the seedling-tree?

MRS. B.

No; but the buds on the graft have attained a state of vigour and perfection which enables them to produce seed; and the seedling-tree may be considered merely as the channel by means of which nourishment is conveyed to them, until age has given it sufficicient vigour to produce fruit-buds of its own. Grafting increases the size of fruits at the expense of the seeds: the rose acacia, when not grafted, bears seeds; when grafted, it bears none, but its blossom is much finer.

Grafting sometimes produces a change of flavour, and generally retards vegetation: it is often employed as a means to retard that of trees, which bud so early in the spring as to be in danger of suffering from the frost. The walnut, for instance, buds a full fortnight later when grafted on walnut.

In regard to the mechanical part of the process, care must be taken to fasten the graft to the tree with soft ligatures, and in such a manner that the vessels of the respective barks may come into contact; then, in order to prevent the extravasation of cambium, the wound must be well covered over with a ball, which is generally made of cowdung and stiff clay. The composition which M. De Candolle recommends for this purpose consists of one pound of cow-dung, half a pound of pitch, and half a pound of yellow wax.

The season for grafting is either in the spring during the ascent of the sap, or in the autumn for the sap of the following spring.

CAROLINE.

But does not the sap rise constantly throughout the summer?

MRS. B.

For the purposes of general vegetation it does; but you must recollect that all germs and buds, are fed by sap, elaborated by vessels appropriated exclusively for that purpose; advantage must therefore be taken of the period when this sap is in full flow, to effect the junction of the two plants.

M. Chondi has distinguished himself by the numerous experiments he has made in the art of

grafting: he divides the plants susceptible of
undergoing this operation into three classes :

1. The *Unitiges,* or plants having one central and
vertical stem, such as firs, larches, and most evergreens.   In these it is the stem which must be
grafted, and not the lateral branches.

2. The *Omnitiges,* every branch of which is
equally a stem, and therefore each is capable of
being grafted.

3. The *Multitiges,* plants in which some
branches are stems and susceptible of being
grafted, and others are not so.

M. De Candolle once saw every branch of a
large pear-tree grafted : this was done in order to
preserve a great number of grafts, which had just
been received from a foreign country.   The following year they were each transferred to separate
trees, and succeeded extremely well.

EMILY.

And how are grafts, when brought from distant
countries, preserved alive ?

MRS. B.

Frequently by dipping them into honey, which,
by preventing evaporation, preserves the internal
moisture.   Another mode is to bury the cut end
of the graft in a moist root, such as a carrot or a
turnip.

It would be very difficult for me to explain all
the various manners of grafting, there being above

a hundred, which may be divided into three classes.

1. Grafting by approach. You bring together two branches of two neighbouring trees, and, cutting off the extremities of each, you graft them together. If three trees be united together in this manner, the stem of the central one may be cut down, and the head will be kept alive and nourished by its two neighbours.

CAROLINE.

This must be a safe mode of grafting rare and delicate plants, as it is attended with no risk; for suppose the junction does not take place, each branch remains uninjured and grows separately, so that nothing is lost.

MRS. B.

It is a good mode, also, for common plants, being so easy and rapid in its results.

CAROLINE.

And might not in a similar manner several stems be united to a single head?

MRS. B.

Yes; if a circle of stems of young poplars be bent and united in a centre, they will form but one head, which, nourished by the surrounding stems, will grow to an enormous size.

Some trees graft themselves spontaneously. If two branches of *hornbeam* happen to grow so close together as to rub against each other when moved by the wind, the outer bark will be worn away by the friction, and the vessels of the two libers will come in contact, which is sufficient to produce a graft. The mere act of two branches growing contiguous, when confined for space, will wound the bark; and, indeed, by whatever chance the vessels of two branches are brought together, in such a manner that the sap can flow from the one into the other, a graft takes place.

EMILY.

You sometimes see a young tree growing from the trunk of an old one of a different kind: is this the result of a natural graft?

MRS. B.

No; it is that of a seed which has sown itself, or of a slip which has planted itself in the hollow of a decayed tree, the rotten wood producing a soil which consists wholly of the richest nourishment. I have seen a fine young cherry-tree grow out of the hollow trunk of an oak, and a vine spring from the old stump of a willow.

EMILY.

You sometimes see two leaves and two flowers cohering together: is not this owing to a natural graft?

Yes; this species of grafting extends also to fruits: the double cherry is thus grafted.

Pray, are roots susceptible of being thus grafted?

It has been so affirmed, but it may probably be an error, arising from confounding subterraneous branches with roots.

The second mode of grafting, and which is in most common use, is by scions: a young branch is cut off from one plant, and grafted on another. In whatever manner the section be made in the graft, one of a corresponding form must be made into the subject, in order that they may fit into each other, and the vessels of the liber come into contact; the union then takes place in the course of a few days.

The mode of grafting by approach bears a considerable analogy to the propagation by layers; and that by scions, to the propagation by slips or cuttings.

True; and the third class, which is grafting *bourgeons* or buds, may in some respects be compared to propagation by seed: it consists in transplanting a bud from one plant to another. For this purpose the bud must be separated from

the parent-plant, in such a manner as to be surrounded by a small disk or shield of bark; for
since the union is effected by the vessels of the
bark, it is through its intervention alone that the
bud can be grafted. A small piece of the alburnum is sometimes cut off in addition to the bark,
but I believe this to be an unnecessary precaution.
The bud may be adjusted on any part of the bark
of the tree, but it is more sure to succeed, if it be
grafted on a spot where another bud had previously existed, in order that the vessels which
conducted nourishment into the original bud, may
pour it into that which is substituted in its place.
In this manner Mr. Tschudy has succeeded in
grafting herbaceous plants, — the melon, for instance, on the gourd; and potatoes on tomatas;
but one herb cannot be grafted upon another.
This species of grafting is rather novel, it being
formerly supposed that herbaceous plants were
not susceptible of being grafted.

The embryos of buds, before they begin to be
developed, you may recollect, are called eyes; and
the graft may be made either when the eye is said
to be sleeping or waking: that is to say, either
in autumn, when the eye is closed for its long
winter-night of repose; or in spring, when it is
open for its summer-day of activity.

Another mode of grafting buds is by transplanting a broad ring or flute of bark, containing
several eyes, and substituting it in the place of a
similar ring cut away from the stock: this is less

sure of success, on account of the number of buds to be nourished.

These several modes of propagation by layers, by slips, and by grafts, are all calculated to improve the fruit; the grand source of the multiplication of plants is the seed, which we shall enter upon at our next interview.

CONVERSATION XVII.

ON THE MULTIPLICATION OF PLANTS BY SEED. —
THE FLOWER.

MRS. B.

WE have now reached that part of our subject
with which you thought it would have been proper
to have commenced — the history of the seed. It
will be necessary to introduce it by a description
of the organs whose office it is to prepare this important part of the plant.

CAROLINE.

That is to say, the flower, which forms the principal part of the study of botanists in general, and
which we have hitherto totally neglected.

MRS. B.

If I have allowed the most beautiful part of the
vegetable creation to remain so long unnoticed, it
was in order that, when I described it, your interest might be excited, not merely by the brilliancy
of its colours, the elegance of its form, or the

sweetness of its perfume, but that, having acquired some previous knowledge of the economy of vegetation, and become acquainted with the essential part it performs among the works of Nature, you would take a deeper and more rational interest both in the blossom and the seed, and that your admiration would be excited by learning, that the most beautiful part of the vegetable kingdom, prepares and ushers into life, that which is most useful. No child has so richly ornamented a cradle as the seed when reposing within the recesses of the flower.

The flower consists of several parts.

The *calyx*, or flower-cup, forms the external integument which shelters and protects the bud before it expands : it consists of several parts, called sepales, resembling small leaves, both in form and colour; and probably performs similar functions, being furnished with stomas. These sepales are, in general, more or less soldered together, sometimes so completely as to form a cup apparently of one piece : hence the calyx has acquired the name of flower-cup.

CAROLINE.

I see that you persevere in deriving every organ of a plant from the budding of leaves.

MRS. B.

When you are a little more acquainted with

plants, I think that you will concur with me in this opinion.

Above the *calyx* rises the *corolla*, which is the coloured part of the flower. It is composed of several petals, either distinct and separate, or cohering so as to form a corolla of one single piece : in the latter case the flower is called monopetalous, though the petals are never originally simple, as this name would seem to imply, but, like the calyx, derive their origin from a circle or *whorl* of leaves. When the petals first burst from the calyx, and expand in all their beauty, they still serve to protect the central parts of the flower : they are at first curved inwards, forming a concavity around the delicate organs which occupy the centre, which not only shelters them from external injury, but reflects the sun's rays upon them like a concave mirror, thus rearing them as it were in a hothouse. When they are full grown, the artificial heat being no longer necessary, and the admission of light and air not only safe but advantageous, the petals expand, leaving the internal organs exposed to the free agency of these elements.

At the base of the petals is generally situated the *nectary*, so called from its secreting a sweet fluid, which has been dignified by the name of *nectar*. This is the store whence the bee derives honey : it affords also abundant provision for the less provident insect tribe, who, rioting in these sweets during a summer, scarcely outlive the fall of

the blossom.   These thoughtless beings, however
unwittingly, act a useful part in the economy of
Nature, which I shall presently explain to you.
The nectar exists in almost all flowers, but is not
always contained in a distinct organ.

CAROLINE.

I have often sucked it from the tubular orifice
of the petals of the honeysuckle.

MRS. B.

That flower produces a great quantity of honey,
and part of it lodges in the elongated tube whence
you suck it.

The most important parts of the flower are those
delicate organs which occupy the centre as the
place of greatest security.   It is here that the seed,
which is to propagate the plant, is lodged.   It is
enveloped in a small leaf, which, instead of expanding its beauties to the sun and air, like its
neighbouring petals, folds itself more closely around
the little treasure it is to protect : the edges of the
two opposite halves of the leaf being thus brought
in contact, they unite and grow together, and the
leaf assumes the form of a pod, or vessel, the shape
of which varies according to the manner in which
the leaf was folded when it first budded.

CAROLINE.

And will you not admit that plants have sensibility, Mrs. B., when you see them showing such signs of maternal care for their offspring?

MRS. B.

No, my dear. Were I sufficiently versed in the physiology of plants, I should no doubt be able to show you, that this tender care of the protecting leaf is the natural result of physical laws.

CAROLINE.

Then I am almost tempted to rejoice that you are not learned enough to do so. I cannot help being vexed when I hear facts, so interesting to the feelings, explained away by the dry results of mechanical or chemical laws.

MRS. B.

You are falling into an error very common among half-learned and superficial observers.

When you feel inclined to murmur at the dry results of physical laws, let not your imagination rest there, but raise your mind from these impassive agents, to their Omnipotent Author: you will then consider them as the unerring instruments which his paternal care has provided, to promote and secure the welfare of his creatures.

CAROLINE.

I now understand and perfectly acquiesce in your sentiments. It is very true that the mind, amazed at the wisdom that is displayed in the laws

of Nature, is apt to consider them as a sort of mechanical cause, rather than as mere agents of an
all-wise and sentient Power.

MRS. B.

To return, then, to the flower and the envelope
of the seed, in which, I trust, you will continue to
take some interest, although we have deprived it
of sensibility; — unless in a poetic sense. When this
leaf is closed over the seed, and its edges soldered
together, it is called an *ovary*, or seed-vessel.
From its summit rises a little thread-like stalk,
called a *style*, which, at its extremity, supports a
small spongy substance, denominated *stigma*.
These three parts form a whole, which bears the
name of *carpel*.

EMILY.

Is carpel, then, synonymous with pistil? For I
know that an ovary, with its style and stigma,
constitutes a pistil.

MRS. B.

Pistils are composed, in general, of several carpels, which, in most flowers, are so neatly fitted to
each other, and so closely adhere together, that
they are considered as a single organ, containing
different cavities for seed; but the most accurate
anatomical researches prove that these several
cavities have each its style and stigma, and form
distinct carpels: thus, the blossom of the apple

and the pear have several carpels soldered together.

CAROLINE.

Oh, yes; for when they become fruits they contain several seeds.

MRS. B.

That would afford no proof of the pistil consisting of more than one carpel, which often contains many seeds; but in the apple and pear the seeds or pips are lodged in separate carpels. It is true, however, that a single carpel forms the pistil of some flowers; such, for instance, is the blossom of the cherry, which, you know, has but one seed, the kernel contained within the stone.

In some flowers, the styles and stigmas remain separate, and the ovaries are soldered together: the flower is then said to have two styles with one ovary, containing several cells or cavities for seed; in others, it is the styles which adhere together while the ovaries are detached, and in some few the adhesion takes place only between the stigmas.

Immediately surrounding the pistil are situated the *stamens;* each of which consists of a slender filament supporting a little bag or case called *anther*, filled with pollen, a species of dust or powder. The anthers when ripe burst, and, being more elevated than the stigma, shed their pollen upon it, and the seeds are thus perfected.

Yet I have heard that there are some plants whose flowers have no stamens, and others which have no pistils: in this case, how can the pollen of the stamens fall upon the stigma of the pistils? Nature has, no doubt, provided some resource to overcome the difficulty.

Or, rather, it is a provision she has specially made in favour of another part of the creation. The pollen is sometimes conveyed by winged insects, which, in penetrating, by means of their long and pliant probosces, within the recesses of the corolla, in order to obtain the nectar, cover their downy wings with the pollen.

This unheeded burden they convey to the next flower on which they alight; and, in working their way to the nectary, it is rubbed off and falls on the stigma: this compensation they make, for the honey of which they rob the flower; and they thus unconsciously labour for those plants, which afford them food. Every insect, however ephemeral, every weed, however insignificant, has its part assigned, in the great system of the universe.

In Persia, very few of the palm and date trees under cultivation have stamens, those having pistils being preferred as alone yielding fruit. In the season of flowering, the peasants gather branches of the wild palm-trees whose blossoms contain stamens, and spread them over those which are culti-

vated, in order that the pollen may come in contact with the pistils and fructify the seeds.

There are two remarkable palm-trees in Italy,
which have been celebrated by the Neapolitan
poet, Pontanus: the one, situated at Otranto, has
no stamens; the other at Brindisi, which is about 40
miles distant, has no pistils, consequently neither of
these trees bore seed; but when, after the growth
of many years, they rose superior, not only to all
the trees of the neighbouring forests, but overtopped
all the buildings which intervened, the pollen of
the palm-tree at Brindisium was wafted by the
wind to the pistils of that at Otranto, and, to the
astonishment of every one, the latter bore fruit.

CAROLINE.

How extremely curious!

MRS. B.

Having now completed our examination of the
flower, it will be necessary to bestow some attention on the stalk which supports it. This is called
a *peduncle*, or *pedunculus*. It generally expands a
little at the summit, and forms a common base by
which the several parts of the flower are connected
together. This little expansion is called *torus*,
which signifies a bed.

EMILY.

It is the bed on which the flower reposes; but

it belongs to the stem, and, I believe, forms no part of the flower?

You are quite right: the flower consists of the calyx, the corolla, the nectary, the pistil, and the stamens. If you pluck off these several parts, the torus will remain on the peduncle; but we shall see hereafter, that, though it forms no part of the flower, it sometimes enters into the composition of the fruit.

The peduncle is not always crowned by a flower: it often branches out into a number of smaller flower-stalks called *pedicels*, each of which supports a flower.

When pedicels diverge regularly from the summit of the peduncle, as rays from a centre, it is called an *umbel*, from the resemblance which the pedicels bear to the branches of an umbrella. A second umbel frequently shoots from each pedicel of the first; the umbel is then said to be compound.

I observe that the peduncle expands, so as to form a base for the pedicels which grow from it, and this expansion is surrounded by a little circlet of leaves — probably bracteas?

Yes they are, and are usually called the *involucrum* of the umbel. The base whence the pedicels radiate bears the name of *receptacle*; and it not only serves to support them, but, the sap being accumulated in this expansion, it becomes a reservoir of nourishment, and supplies them with food which they each convey to their respective flowers.

EMILY.

Does not the Laurustinus blossom in this manner? I have often observed that its peduncle spreads out into a number of different ramifications.

MRS. B.

They do not spring from a common centre, and, consequently, can have no common receptacle; but are irregular, like the branches of a tree, and the bunch of flowers they support may be compared to its head. It is hence called a *cyme* or *cyma*.

The peduncle often throws out small pedicels at regular distances, as you may have observed in a bunch of currants: this sort of cluster is called *raceme* or *racemus*, and its flowers open in succession from the bottom to the top.

In some plants the flowers are placed around the peduncle on such very short pedicles that they assume the form of a *spike* or *spica*. When thus disposed they blow in succession, so that those at the bottom of the spike have withered before those at the top are unfolded. Plantain blossoms in this manner. In other plants the flowers are crowded still more closely around the peduncle, and form an ear: such is the mode of flowering of corn and grasses; in others, they grow in clusters or irregular bunches like the vine. In many trees the peduncle assumes the form of a spike, *articulated* with the branch, and covered with the remains of degenerated bracteas, resembling scales, under each of which a flower lies concealed : the hazel, the willow, the alder, and the hornbeam, blossom in this manner.

EMILY.

I have paid so little attention to the manner of flowering of plants, that I was not at all aware they afforded so great a variety.

MRS. B.

I am far from having enumerated them all, for every different mode in which the pedicels diverge from the main stem, and which produces a different arrangement of flowers, bears its own peculiar name ; but the whole is included in the term *inflorescence*, which expresses the various modes in which the stem of a flower is divided, and, consequently, the arrangement of the flowers upon it.

Plants blossom at regular periods ; varying, however, according to the temperature of the country in which they grow, and the vicissitudes of the season. Linnæus formed a register of the season of flowering of different plants, which he called the Calendar of Flora, but no allowance being made for these modifications, it is very imperfect.

EMILY.

I have observed that there are some trees which

regularly blossom earlier than others, of the same species and in the same situation : whence does this arise ?

MRS. B.

It is not ascertained; but as every peculiarity of an individual plant is preserved when it is propagated by layers, slips, or grafts, advantage has been taken of this anomaly to produce early vegetation. Mr. Knight, by carefully selecting those potatoes which first sprouted for replanting, obtained in the course of a few years plantations of potatoes very considerably earlier than the usual season.

In hot climates the fig-tree produces two crops of fruit, and it is in some countries necessary to accelerate the ripening of the first, in order to leave time for the second to come to maturity, in due season. With this view, the peasants in the isles of the Archipelago, where this fruit abounds, bring branches of wild fig-trees in the spring, which they spread over those that are cultivated.

EMILY.

This is, no doubt, the same process as that of the fructification of the palm-trees in Persia.

MRS. B.

It was long supposed to be so ; but it is now ascertained that the cases are quite different, the only use of these wild branches being to serve as a vehicle to a prodigious number of small insects, called cynips, which perforate the figs in order to make a nest for their eggs, and the wound they inflict accelerates the ripening of the fruit nearly three weeks.

EMILY.

Does the insect produce this effect by the injection of some stimulating fluid into the wound it makes, or is it owing to the growth of the eggs it deposits?

MRS. B.

The precocity does not appear to result from either of these causes: it is, indeed, not well known; but I should think may probably result from the punctures of the insects, impeding the free course of the sap, and producing, like the annular section, an accumulation of sap in those parts, which, by affording additional nourishment to the neighbouring buds, accelerate their developement.

Have you not observed that fruits which are worm-eaten ripen earliest?

EMILY.

Yes; but I thought that the worms attacked those which were first ripe.

MRS. B.

I do not allude to the external attacks of worms and insects, but to the maggot born and bred within the fruit; and the nest of eggs, whence it drew its existence, was in all probability the cause of the precocity of the fruit.

Means may also be taken to retard the period of blossoming : too much nourishment is injurious at that season, and sometimes wholly prevents it. Much water is also prejudicial : the water is drained from the rice plantations when the rice is in flower; and the watering of gardens should be diminished.   Snow late in the spring has, in mountainous countries, been known to retard the blossoming of corn till the following year.

It is remarkable that the conveyance of plants from one country to another appears to accelerate the period of flowering; for plants brought from foreign climes blossom earlier than usual, the first year of their emigration.

CAROLINE.

That is very singular.   Can it be owing to the excitement produced by the motion of the carriage ?

EMILY.

May it not rather be attributed to the total cessation of vegetation during the journey, when the plant is confined by packing, and the consequent re-action which takes place on its being replanted ?

MRS. B.

It is a point very difficult to explain.   It frequently happens that, after blossoming, the fruit perishes from debility.   An annular incision of the bark (which you may recollect arrests the cambium in its descent) increases the vigour of the blossoms by affording them more nourishment; but the ring, when made for this purpose, should be very narrow, in order that the upper and under edges of the severed bark may re-unite when this superabundance of food is no longer required in the upper part of the plant. M. Lancris makes the ring of such narrow dimensions, that the separation of the bark, lasts only during the flowering of the plant; at the end of which period, the protuberance at the upper edge of the bark having swelled out, till it reached the lower edge, and being still soft, the contact and gentle friction produced by the continuance of its swelling occasions it to burst: it then amalgamates with the lower edge, when the wound is healed, and the general circulation restored.

CAROLINE.

That is to say, that the upper edge of the bark grafts itself upon the lower edge?

MRS. B.

Precisely so. This operation has been performed on the vine with some success; but these experiments have not been sufficiently extensive for their general results to be relied on. Its effect on fruit-trees, we have already observed, is very precarious: the branches of fruit-trees not being completely lopped every year, like those of the vine in vineyards, they are liable ultimately to suffer from the derangement of the circulation.

It answers better with fruit-trees whose seeds are pippins, such as the apple and the pear, than with such as have stones and kernels, like the peach and the apricot, because, when the incision is made, the latter exude a gummy juice, so that they are liable to lose more than they gain by the operation.

There is another cause which frequently prevents the fruit from being formed. It is when water falls upon the stamens: this makes them burst before the due season, and the pollen, instead of being shed upon the pistil, is lost. Rain, and even heavy mists, the latter of which, still more than the former, insinuates itself into the flower, very frequently produces this effect.

CAROLINE.

But all blossoms are exposed to mists and showers: how then can any fruit be set?

MRS. B.

It is evident that Nature has decorated plants with a much greater number of blossoms, than she designed to convert into fruit, for the plant would have no means of bringing so great a quantity to maturity. Look at an apple or a cherry tree in blossom, and you will observe, that were every flower to produce a fruit, not only would it be impossible for the tree to nourish so great a crop, but even its branches would be unable to sustain them. Therefore, though every shower may destroy, or, rather, prevent the formation of a quantity of fruit, it would require heavy and continued rains to prove fatal to the whole. This, however, sometimes happens, particularly to the vine, which in wine countries is a very serious calamity.

I have still some further observations to make on flowers, but I think you have learnt as much to-day as you can well remember; we will, therefore, reserve what remains to be said on them till we meet again. In the mean time you may refresh your memory on what I have taught you by examining this drawing [Plate I.], in which the various organs of a plant are delineated in the representation of a pea.

This second drawing [Plate II.] represents a plant of the class of Monocotyledons of the liliaceous family, in which it is a disputed point whether the coloured part of the flower is a corolla or a calyx.

EMILY.

It has, surely, much more the appearance of a corolla composed of six petals, than of a calyx consisting of six folioles. One of these two organs, then, is wanting in this family?

MRS. B.

In order to avoid error by deciding which of them it is, botanists call the coloured part of a flower of this description a *perigone*, or *perianth*, composed of one or more pieces; that of the tulip has six.

Plate I

Pub.^d by Longman & C.º June 22, 1829.

Plate II

Pub.ᵈ by Longman & Cᵒ June 22, 1829.

Plate II

Pub.d by Longman & Co June 22, 1829.

CONVERSATION XVIII.

ON COMPOUND FLOWERS.

CAROLINE.

I have been studying your drawings, Mrs. B., and imagined that I understood them perfectly; but when I attempted to make out the several parts on a real flower, I am sorry to say that I found myself quite at a loss.

MRS. B.

What flower did you choose for this purpose, my dear?

CAROLINE.

I was, perhaps, too confident of my powers of discernment, for I selected one that had a totally different appearance from the pea or the tulip: it was a China Aster. I made out a calyx and a corolla, but the rest was all perplexity.

MRS. B.

You have fallen into an error which many botanists have done before you: you took the China Aster for a single flower, whilst it is, in fact, an assemblage of flowers, called in botany a *head.*

Here is a China Aster I have just gathered : let us examine it. [See Plate III.] The stem or peduncle is terminated by what *you* call a flower, and what *I call* a head of flowers. The extremity of the peduncle, you see, expands into a white disk, called a receptacle, analogous to the receptacle of the umbel, and in this all the florets are inserted : it is not only the basis on which they rest, but serves them also as a magazine of food. In the China Aster it is flat, but in many other plants of the same family it is more or less convex : it is sometimes as thin as a sheet of paper, as in the *Scorzonera ;* at others, it is very fleshy, as in the artichoke.

EMILY.

Is it that internal part of the artichoke on which the choke rests, and which is so good to eat ?

MRS. B.

Yes. Around the receptacle of the China Aster you see there are a considerable number of small leaves, or bracteæ.

CAROLINE.

That is what I supposed to be the calyx.

MRS. B.

It is a very natural mistake. It is, indeed, a sort of calyx common to the whole head ; but as each floret has its separate calyx (and a calyx of so peculiar a nature as not to be overlooked), this com-

Plate III

Pub.d by Longman & Co June 22, 1829.

mon calyx is distinguished by the name of involucrum, analogous to the involucrum which surrounds the *receptacle* of the umbel.

EMILY.

There seems to be a considerable resemblance
between the umbel and the head of flowers?

MRS. B.

That is very true. If you conceived the branches
of an umbel to be so extremely short that they
could not be distinguished, the umbel would be
similar to a head; and this is exactly the case of
the *Eryngiums.*

EMILY.

The involucrum of the China Aster differs, however, in one respect from that of an umbel, the
bracteas of which it is composed being much more
numerous, and disposed in several rings or whorls
around the stem.

MRS. B.

That is the case with the greater number of
compound flowers, but it is not universal; for in
the Salsafy (*Tragopogon*), the *Orthonna,* and several
others, the bracteas, or, as they are more commonly called, the folioles of the involucrum, are
placed in a single ring. When disposed in several
rings, they are sometimes equal; at others, vary in
size: they are sometimes curled up; at others,
spread out. Some are soft, others scaly; and there
are some which terminate in a species of thorn or
prickle, as in the thistle. These varieties in the

nature of the folioles serve to distinguish the numerous class of *heads of flowers*, which constitute
no less than one-twelfth part of the vegetable
kingdom.

It is then, indeed, very necessary to make acquaintance with so numerous a body of plants.
But you have mentioned both compound flowers,
and flowers growing in a head : — are these terms
synonymous?

Certainly not; and I am obliged to you for re-
minding me of a want of accuracy.

All flowers which shoot in numbers from a common receptacle, either flat, or slightly elongated,
are called *heads.*

The flowers, analogous to leaves without petioles, are called sessile; such are the Scabiosa, and
many others.

Now, amongst this extensive class, there is one
family distinguished from all the rest by the cohesion of their anthers, so as to form a tube around
the style, and it is this peculiarity which constitutes
the *compound flower*, or family of *Syngenesia ;* and it
is to this family that I shall more particularly direct your attention, as the China Aster, which we
are examining, belongs to it.

You see all these little yellow parts in the centre
of the head, and these violet leaves which spread
out around it, and which you took for petals: they
are all of them distinct flowers.

EMILY.

Is it possible ! Such a concourse of tiny flowers, so closely crowded together in the centre; and these appear totally different from the violet-leaves, which you also call flowers.

MRS. B.

They are far from being so different in their structure as you would imagine from their appearance. In the China Aster, and in several other of the Syngenesia, the florets, though distinct, are not separated from each other by any intervening body; but there are some plants of this family, such as the endive, the artichoke, and the camomile, whose florets are separated by a species of small bracteæ, which have been called palix or *chaff*, and which shoot up from beneath each floret. These bracteæ are sometimes of a scaly nature, and sometimes they assume the appearance of bristles or hairs. The choke of the artichoke, before the blossom is developed, is of this description.

EMILY.

The artichoke, then, is a compound flower; and the only part of this plant that I am unacquainted with is its blossom, which is not developed when it is served at table. We there eat the receptacle and the most tender part of the leaves which compose the involucre, whilst the choke we carefully extract in order to avoid eating it.

I advise you, when an opportunity occurs, to make acquaintance with the blossom. Let us now return to the China Aster: what I have hitherto told you relates more to the mode of flowering; but we will examine the structure of the flowers themselves. [Plate III.] Look at one of those little yellow florets in the centre of the head: with the assistance of this magnifying-glass you will be able to follow me as I describe the different parts. Observe, first, this white spot, which forms the basis of the floret: it consists of the tube of the calyx, and contains the ovary or seed-vessel to which it adheres.

The external part of this tiny tube, then, is the calyx, and the internal part the ovary; but what are those little hairs which crown the tube, and grow from either the calyx or the ovary, I know not which?

They proceed from the margin of the calyx, which assumes this singular appearance, because its natural growth, in the form of sepals or leaves, is impeded by the pressure of the adjacent florets.

And I conceive that the calyx may be stinted

in its growth for want of food, as well as for want of room.

MRS. B.

That may also produce some effect in checking its growth; but the elongated form, which the edges of the calyx assume, must be chiefly owing to pressure. In some heads, in which the florets are not so crowded as in the China Aster, the calyx wears a more natural appearance, being shaped like a cup, and is of a membranaceous texture; in others it resembles small scales: in the present instance, and in most compound flowers, it consists of a species of hairs, either separate or glued together. It was formerly considered by botanists rather as an appendage to the calyx than forming a part of it, and was distinguished by the name of tuft or *pappus*; and though this name applies literally only to hairs, it has been extended by analogy to all the various forms which this organ is capable of assuming.

CAROLINE.

But this little feathery tuft appears much more ornamental than useful: it cannot, I think, in such a form, afford protection to the flower.

MRS. B.

The florets, being so close together, protect each other: the use of the tuft is to assist the fruit to disengage itself from the involucrum, and then to transport it to a distance; for the pappus

D 3

remains upon the fruit after the blossom has
fallen. There are some few compound flowers
the calyx of which is not at all elongated, and
which, consequently, have no pappus.

EMILY.

How, then, does the fruit or seed disengage
itself from the involucrum?

MRS. B.

When the tuft is wanting, the fruit is furnished
with other means of separating itself from the
parent-plant. Sometimes the receptacle rises up
after the blossom is over, to force out the fruit; at
others, the weight of the head, when it is mature,
bends the pedunculus, and the seeds fall to the
ground. Thus you see that every difficulty is
foreseen and obviated in the admirable structure
of which I am endeavouring to give you a mere
outline.

CAROLINE.

You speak sometimes of the fruit, and sometimes of the seed, which the tuft wafts away : do
you mean to use these terms indifferently, or have
they each of them a distinct meaning?

MRS. B.

I was not quite correct in so expressing myself;
but the error was very trifling, as you will perceive, when I have explained the difference to you.
I said that the small body, to which the tuft was

attached, was composed of the ovary and the calyx. The ovary contains a single seed, which has its own particular covering, called Spermoderm. Thus the single seed of each little flower is enveloped in three integuments, adhering to each other — the calyx, the ovary, and the spermoderme. In some compound flowers, these three integuments are distinctly seen; but in others they are only supposed to exist by analogy, without being actually visible. These three integuments, soldered together, form a peculiar species of fruit, which was formerly called a naked *seed*, but is now distinguished by the name of *Achenium*. In some families (such, for instance, as the *Epilobiums* and the *Apocinums*), small feathery tufts grow within the germen, and are attached to the seed. These tufts bear the name of *Coma*, and serve the same purpose as the pappus, though their origin is different. But we are digressing from our China Aster.

EMILY.

Pray let us return to it; for it has become very interesting, since I have learnt how much there is in these little things which you call florets.

MRS. B.

Above the ovary, and within the pappus, you may perceive a yellow tube, terminated by five small teeth; this is the corolla. If you slit it up with the point of a penknife, and look very close, you will see five little stamina: they have each their filaments, which appear to grow out of the tube of the corolla, and each of these filaments is terminated by an anther.

CAROLINE.

I can see only one anther.

MRS. B.

Because the five anthers adhere together, so as to form a cylindrical tube, through which passes the style, the extremity of which spreads out into two small branches. It is this tube which constitutes a characteristic distinction of the compound flower. The five anthers, of which this tube is composed, open internally by two small slits. The style is also furnished with a peculiar species of stiff hairs, called *sweeping hairs;* because they are designed to sweep the pollen from off the anthers, so as to make it fall upon the stigma.

EMILY.

How wonderful that a little yellow atom, which I hardly looked at, should contain so great a variety of curious organs!

MRS. B.

The more you study nature, the more beautiful and magnificent it will appear. But we have not yet done with the China Aster. You understand the structure of the yellow central florets; but the purple ones, which form the circumference, are very different.

In appearance, certainly, they are ; for they look exactly like long narrow flat leaves.

Pull out one of these violet leaves, and you will see that the extremity, by which it is attached to the receptacle, is not flat, but round and hollow, in the form of a tube. Suppose it to be an elongated tube, slit open lengthwise and spread out, and you will form a tolerable idea of this species of corolla. These are called *ligulate florets ; ligula* being Latin for a strap : hence they frequently bear the name of strap-shaped florets.

At the upper extremity there are five small teeth, like those of the yellow tubes.

But here, Mrs. B., is another China Aster, in which there are no yellow florets ; the head is entirely composed of these flat violet florets, which we took for petals.

This is a double China Aster. All its florets have undergone the change which in general takes

place only with those situated at the circumference
of the head; and you have here a proof of the
two sorts of florets being of the same nature, since
they are susceptible of being transformed from
the one into the other. If you examine the flower
attentively, you will see that the ligulate florets
have no stamens, and even the style often appears
imperfect; so that florets of this nature yield no
seed, and when a head is entirely composed of
them, it is incapable of propagation.

CAROLINE.

Yet here is a scorzonera which has only flowers
of this description, and it produces seed?

MRS. B.

I will explain to you whence this difference
arises. Compound flowers exist in three different
states: — the head is sometimes composed entirely of tubular florets; the artichoke and the
thistle are of this description: they are called flosculous, and, with some few peculiar exceptions, all
the florets yield seed. A second state is when the
head is composed entirely of ligulate florets, having
stamens and styles; such as the double China Aster,
you have just observed: these form the class called
Ligulate, or, as they are sometimes, though less properly, called, *Semi flosculous.* The scarzonere and
the endive belong to this class, and all the florets
yield seed. In the third state, the florets in the
centre of the head are tubular, and those at the

circumference flat or ligulate: these are denominated *Radiate*. The dahlia, the aster (including the
China Aster), the camomile, the daisy, and many
other plants, are comprehended in this division.
The central florets generally yield seed, while the
lateral ones are barren.　These varieties of structure, combined with those which exist in the
receptacle, the involucrum, and the pappus, have
enabled botanists to separate and divide into classes
the numerous compound plants which are spread
over the face of the globe.　There is also a fourth
state of compound flowers, in which the corolla is
divided into two lips; they are called *Labiate florets*.
But I shall not enter into any details on this class,
as it is found only in America, and is very rare in
the botanical gardens of Europe.

EMILY.

I am glad that you spare my memory; for I fear
I shall have some trouble to recollect all you
have taught me concerning those which grow in
Europe.

MRS. B.

I assure you that I have endeavoured to make
the subject as easy as I possibly could, and have
omitted many difficult parts; but in this case, as
in every branch of botany, you will understand
clearly, only by seeing with your own eyes.　Ana-
lyse the compound flowers you meet with; and
when you have examined a few, you will comprehend them better than all my explanations will

enable you to do. I do not pretend to make you adepts in botany; I merely wish to direct your attention to the observation of the works of Nature: they will speak for themselves, and in a language far more eloquent than I possess.

61

CONVERSATION XIX.

ON FRUIT.

MRS. B.

It is now time for us to take leave of flowers,
and turn our attention to the fruits which they
produce; in which state the seed may be considered as entering into a second stage of existence.

After the flower has performed its office of
fructifying the seed ; the petals, and every organ
which is not destined to become a part of the fruit,
wither, and fall off.   In the mean time, the ovary
grows, and gradually assumes the appearance of a
fruit.

EMILY.

Is the fruit, then, formed from the original little
leaf which so carefully guarded the seed when the
flower was in blossom, and which you called a
carpel ?

MRS. B.

Yes; when it assumes the form of fruit, it is
frequently called by botanists *pericarp.*

62 ON FRUIT.

CAROLINE.

But I suppose it retains the name of seed-vessel, since it contains the seed in the fruit as well as in the flower?

MRS. B.

Certainly. Now let us take, for example, one in which the form of the original leaf is not wholly obliterated — this pod of a pea, for instance: you may plainly see that it consists of a leaf doubled over the seeds, with its edges united.

EMILY.

This pod, which is very young, is almost flat; but here is a larger one, which is become convex, in order to make room for the growth of the peas; and I perceive that the older it grows, the more it loses the form and appearance of a leaf.

CAROLINE.

In shelling peas, I have observed that the pod readily opens where the edges of the leaf have been soldered together; but if you attempt to sever the pod at the opposite seam, which I suppose forms the midrib of the leaf, it is much more difficult.

MRS. B.

You are mistaken there, my dear; for, in shelling peas, the pod is opened by splitting asunder the midrib of the leaf. When the pod is ripe, this rib opens of itself, and the opposite suture or seam, formed by the soldering of the edges of the leaf, also gives way; so that the pod is separated into two halves or valves, and the seeds detach themselves and fall to the ground. This is a natural mode of opening, for the purpose of shedding their seed, which is common to a great number of pericarps: it is called *dehiscence*.

CAROLINE.

Then the peas, which I thought had been attached to the midrib of the leaf, must grow from its margin: that seems very singular. Is there any instance of leaves, in their common state, bearing seeds thus?

MRS. B.

Yes; a leaf has been discovered (the *Bryophyllum*) which has this extraordinary property: it bears germs, susceptible of becoming young plants, and these are situated on its margin, like the seeds in a carpel. The same structure is found in the *Malaxis paludosa,* a little plant, growing occasionally in bogs in this country.

CAROLINE.

Well, it is not very difficult to comprehend that a leaf may be converted into a pod, which you botanists dignify with the name of fruit; but I cannot conceive how you can metamorphose a leaf into what we ignorant people call fruits; such as an apple, a cherry, or a plum.

With a little farther explanation, I hope I shall
be able to accomplish this. Do you recollect the
structure of a leaf?

It has a smooth upper surface, and an under
surface more porous, of a rougher texture, and
generally downy or hairy. Between these surfaces
lies the pabulum, a softer body, consisting of an
expansion of the cellular system, and this is tra-
versed and intersected by the fibrous vessels which
form the ribs of the leaf.

Extremely well. In the pea-pod these several
parts are distinguishable. The leaf is doubled
upon its upper surface, so as to render the under
surface external.

The most porous surface must, of course, form
the outside of the pod, otherwise the stomas could
be of no use.

This is the case not only with the pea but with
all carpels. The external surface takes the name
of *epicarp:* — the upper surface of the leaf forms
this thin delicate skin, which lines the interior of
the pod: it is called *endocarp;* and the pabulum
of the leaf is this soft intermediate layer, which
is denominated *mesocarp.*

CAROLINE.

Oh, what hard words to remember, Mrs. B. !

MRS. B.

You will, perhaps, be able to retain them more easily if I explain their derivation: *carpos* is the Greek word for fruit, and *epi* for upon or over.

CAROLINE.

That clears up the whole difficulty: for it is easy to understand that *epicarp* signifies the outside skin which is *upon* the fruit; *endocarp* the inside skin; and *mesocarp*, no doubt, means the middle substance between the two. Now, if you will be so good as to tell me the derivation of the word *peri*, I shall not forget the meaning of pericarp.

MRS. B.

Peri signifies about or around; so pericarp means about or around the fruit. According to this definition, the seed alone is considered as the fruit; but, in the usual acceptation of botanists, the pericarp itself constitutes the principal part of the fruit.

A leaf, forming a carpel or pericarp, may be folded in a variety of ways, either cylindrical, or like a cornucopia, or doubled a little convex like a pod; but, however diversified the form of the fruit, it results always from the manner in which the leaf was originally folded, when it first budded.

Now, into what sort of fruit, do you wish that I should convert one of these pericarps ?

You speak with the same confidence, Mrs. B., as if you were going to perform the metamorphosis with a fairy's wand; and make me expect to see it accomplished, with the same facility that the pumpkin was converted into a coach for Cinderella. However, I shall endeavour to increase the difficulty of your task, by making choice of a fruit which bears no kind of resemblance to a leaf — a peach, for instance. Will it not require the utmost effort of your art to effect this transformation?

Far from it; for the peach is one of the most simple of fruits : it resembles the pea-pod, in being composed of a single carpel, but it is still less complicated, for the carpel contains but one seed — the kernel within the stone. The skin is the epicarp. Do you not recognise the hairy cuticle of the under surface of the leaf in the downy skin of the peach ? Then the cellular texture of the pabulum, absorbing a great quantity of sap, and swelling out as it grows, forms the fleshy substance of the fruit : — this is the *mesocarp*. Finally, the upper surface of the leaf being, in a great measure, deprived of moisture, and starved, as it were, by the voracious appetite of the mesocarp, its fibres contract, become tough, then indurated, and are at length converted into a shell or hollow stone, which affords most secure shelter for the seed : — this is the *endocarp*.

CAROLINE.

What a very curious transformation! Every vestige of the ribs of the leaf is obliterated in the fruit; but traces of contraction of the endocarp are discernible in the seams and wrinkles with which the stone is covered.

EMILY.

There are also indications of its being composed of two valves, for a sharp instrument will split it open, and divide it into two parts, and, when it is diseased, it separates of itself. Then the curved indenture, which runs along the peach on one side, I think, points out the seam of the carpel. And pray, Mrs. B., are all stone-fruits formed in the same manner?

MRS. B.

Yes, they are. This class of fruits is distinguished by the name of *Drupe* or *Drupa:* among these you will, perhaps, be surprised to hear, that the almond and the cocoa-nut are classed.

EMILY.

They certainly bear very little apparent resemblance to fleshy stone-fruits, being wholly destitute of a fleshy mesocarp.

MRS. B.

In these dry drupes, the mesocarp assumes the form and texture of coarse thready fibres, which form the external covering of the nut: the endocarp is the hard woody nut, and the smooth skin with which it is covered is the epicarp.

Who would ever have imagined that the flesh of the peach, so delicate and luscious, and the coarse fibres which enclose the almond, had both a similar origin! I suppose, then, that it is the almond which absorbs the chief part of the nourishment, for the whole of the pericarp is dry and meagre.

Were fruits not so treacherous in their appearance, I should conclude that the apple and the pear derived their fleshy substance, like the peach, from the swelling out of the mesocarp; but I have so often been mistaken in my conjectures, that I make the enquiry with diffidence?

You are right not to be too confident; for the apple and pear are quite of a different description from the drupe. But do not let us proceed too fast, and by degrees I hope I shall be able to make you comprehend them all. I began by selecting the most simple cases, in order to be well understood: we must go on upon the same plan; for in natural science we cannot, as in chemistry, make experiments which gradually lead us from the simplest to the most compound combinations; but Nature makes these experiments for us, and our business is only to arrange the combinations she exhibits in methodical order. We have hitherto considered fruits formed of a single carpel; but it is not difficult to conceive that a fruit may be composed of several carpels. Take, for example, this Pœony: it consists of a number of carpels, each of which exactly resembles a pod. You recollect my telling you that the pistil of a flower was commonly formed of several carpels: such flowers will produce fruits with a similar number of carpels.

CAROLINE.

Oh, yes; and I recollect your saying that this was the case with the apple and the pear.

MRS. B.

A little more patience: we are not yet arrived at the apple and the pear.

EMILY.

But cannot you show us some other fruits similar to that of the pœony? for I see it is only by natural specimens that one can understand the curious transformation of flowers into fruits.

MRS. B.

It is very true, that it is necessary to observe the flower in order to understand the conformation of the fruit. Here is an *apocynum:* its flower bears two carpels, which differ but little from a pod : these carpels are distinguished by the name of *follicles.* Here is another example : it is a variety of the cherry, which, instead of bearing a single drupe like the common cherry, bears several.

We will now proceed another step, and examine the raspberry : this fruit consists of a considerable number of small fleshy carpels, all of which result from a single flower. It resembles an aggregation of small drupes.

EMILY.

The number of carpels, then, I see offers no difficulty : a flower may bear one carpel, like the pea; or two, like the apocynum; or five, like the pœony; or a still greater number, like the raspberry.

MRS. B.

Now, examine this long narrow pericarp of the wall-flower : of what number of carpels do you suppose it consists ?

CAROLINE.

Of only one, for it is a pod similar to that of the pea or the bean ; — but no, on opening it, I perceive that a thin partition runs down the middle, which divides it into two cavities, and that there is a row of seeds in each. This pod must therefore consist of two carpels growing together, so as to form but one fruit.

Pericarps of this desciption are called *siliques*, not pods. The pod belongs to the *leguminous*, the silique to the *cruciform* family.

We have seen many instances of the organs of flowers being soldered together: the petals, for instance, are frequently united so as to form a corolla of a single piece; the stamens often cohere together by their filaments; the anthers are united so as to form a tube in compound flowers; and it is not more difficult to conceive that several carpels should be soldered together, and form fruits, having different cavities or cells for seed.

You may consider it as a law of Nature, that the number of cells for seed contained in a fruit, implies the number of carpels soldered together in its formation. This law, however, admits of exceptions, which require some further explanation. The carpels, you allow, consist of folded leaves: if these reach to the centre of the fruit, the cells will be complete; if they reach but half way, the centre will be hollow and empty; for the partitions formed by the folding of the leaves will only reach half way; of this the poppy is an instance. If the leaves be still less folded, they will spread out in growing, and the fruit, though

composed of several carpels, will only have one cell:
the melon is an example of this kind.

How, then, can you distinguish a fruit that has
but one cell for seed, because it is formed but of
one carpel, from a fruit that has but one cell,
though originally composed of several carpels ?

I think I can explain that. When the fruit
consists of one carpel only, the seed will be situated in a row on one side of the carpel; but when
it consists of several carpels, there will be as many
rows of seeds as there are carpels, since each carpel bears its row of seeds.

You are right; but recollect that each row,
though apparently single, is in fact a double row,
the seeds being attached alternately to each valve
of the carpel.

Well, now that you have seen and understood
the result of the soldering of carpels together, and
the effect of the leaves, of which they are composed,
being more or less folded or curved inwards, you
may readily conceive that such differences are sufficient to account for the various forms of fruits.

When the carpels are *verticillate*, that is to say,
situated around a common axis, or a little column
(called *columella*), the division of the carpels often

completely disappear externally, and the fruit assumes a spherical appearance ; but if the convexity
of the carpels be greater than that of the whole
fruit, each carpel protrudes externally, forming a
rib, such as those of the house-leek.

EMILY.

The shrub, which is dignified with the name of
Pæony-tree, has the carpels of its fruit enclosed in
a sort of membrane, which covers them completely.

MRS. B.

This membrane, according to the celebrated
Mr. Brown, appears to be a prolongation of the
Torus, or base of the stamens, which grows over
the carpels, and, in some instances, adheres to
them ; but this species of conformation is very
rare. One that is much more common, but also
more complicated, is when the carpels not only
cohere together, but are also soldered with the
calyx; so that when the blossom falls, the fruit
which grows is composed of the carpels and the
calyx, forming a single body.

EMILY.

This must produce a fruit of a very singular
appearance.

MRS. B.

Not so much so as you imagine; for the apple
and the pear are of this description. This mode of
growing can be easily understood when the fruit is

traced from its primitive existence in the flower; but I can give you an infallible test to know whether the fruit, when already grown, is of this description. You see the eye at the top of this pear : it is formed by the remnants of the sepals, or leaves of the calyx ; and whenever you see such an eye at the summit of a fruit, you may be assured that it con- sists of the carpel and the calyx soldered together. All fruits whose seeds are pips, are of this nature, and are distinguished by the name of *Pome.*

EMILY.

The quince, I am sure, then, consists of the calyx soldered to the carpels, for it has a very large eye: but is the medlar also of this descrip- tion ? — it has a considerable opening at the top, somewhat resembling an eye.

MRS. B.

Yes; and the aperture results from the calyx not completely covering the carpels : these, there- fore, are visible between the teeth, or indentures, which terminate the calyx.

EMILY.

I see that the metamorphosis of a flower into a fruit is in many cases a very complicated affair, and not so easy to understand as I had imagined from your first explanation. Is it the calyx which forms the skin, and the pericarp the flesh of the pome ?

MRS. B.

It is difficult to distinguish these organs, when cohering together. The calyx, however, being external, must naturally form the skin of the pome; part of it may also enter into the composition of the flesh, together with a portion of the pericarp. The heart, or core of the pome, consists of the endocarps of five carpels, each containing two seeds or pips.

EMILY.

The orange has no eye; otherwise I should have thought it had been a fruit of a similar construction.

MRS. B.

Far from it: the orange is a pulpy, not a fleshy fruit, like the *pome* or the *drupe*. Now pulp does not, like flesh, result from the growth of the mesocarp, but is a peculiar succulent substance, situated inside of the carpels: those of the orange consist of the quarters into which the fruit may be easily divided when the rind is peeled off, and the seeds are imbedded in the pulp contained within them.

CAROLINE.

True: they are not lodged in a core, like the apple or the pear; nor in a shell or nut, like the peach or the plum. But might not a fruit have both flesh and pulp?

MRS. B.

Yes; the quince is an instance of this combination: the flesh is, like that of the pear, situated

E 2

outside the core or cells containing the seeds, and within those cells the quince contains pulp; but this species of complication is not common.

CAROLINE.

And pray, under what head do you class those fruits in which the seeds are promiscuously situated, such as the gooseberry, the currant, and the grape?

MRS. B.

They are distinguished by the name of *Bacca*, or berry: in these the mesocarp is soft and succulent. Although the seeds are attached to the endocarp, yet the latter is obliterated when the fruit is ripe. A strawberry is not properly so called, because it does not belong to the class of berries. It consists of a fleshy substance, formed by the expansion of the summit of the pedunculus, in which the several parts of the flower are inserted, and which we have called the *torus*. The small grains which you see upon its surface are so many little carpels, each of which contains a seed.

CAROLINE.

They are so small and dry that they look like naked seeds.

MRS. B.

The pericarp fits closely to the seed, so that they seem to form but one body; but they may, thus united, be considered as so many distinct little fruits, imbedded in the soft substance of the torus.

CAROLINE.

They would be very little appreciated as such, were it not for the delicate flavour of this soft substance.

MRS. B.

These little grains, though dry, are analogous to the small fleshy spherical bodies which form the raspberry; and the white conical substance which remains upon the calyx of the raspberry, after the fruit is pulled off, is analogous to the fleshy substance of the strawberry; for they both result from the growth of the torus.

CAROLINE.

With this difference: in the one it is the torus; in the other, the berry, or true fruit, which is good to eat.

EMILY.

And is not the conical expansion of the torus, on which the raspberry grows, analogous to the axis or stalk which traverses the mulberry? For these two fruits bear a great resemblance to each other.

MRS. B.

You are falling into an error to which every one is liable who judges from the appearance of the fruit without having previously studied the flower. If you examine the blossom of the mulberry, you will see that it consists of several small sessile florets disposed around the axis; that each of these, after the blossom has fallen, forms a dis-

E 3

tinct fruit, consisting of the carpel and the calyx:
these fruits being fleshy, and situated so near to
each other as to come in contact in growing,
cohere together; so that a mulberry, which is
in fact an aggregation of several different fruits,
proceeding from as many different flowers, wears
the same appearance as a raspberry, which is the
result of different carpels belonging to the same
flower.

EMILY.

There are, then, if I mistake not, no less than
four degrees of complication in the composition of
a fruit.

First. Fruits formed by a single carpel, such as
the pea or the peach.

Secondly. Those formed by several carpels, the
produce of a single flower, like the pæony and the
raspberry.

Thirdly. Those formed of several carpels, surrounded by and soldered with the calyx, such as
the apple and the pear.

Fourthly. Those formed by the aggregation of
several fruits produced by different flowers.

MRS. B.

Your enumeration is perfectly correct. I will
give you some further examples of the latter
description. The cone of a pine or fir tree consists of an aggregation of fruits, produced by as
many different flowers, having each a single seed:

these flowers are separated from each other by
bracteæ, which remain after flowering: they grow
tough and hard, and enclose each of the fruits
as it were in a case, the aggregation of which
forms the fir-cone.

EMILY.

This is a kind of fruit quite new to us; and the
cone of the magnolia is, I suppose, of the same
description.

MRS. B.

No; the cone of the magnolia proceeds from
several carpels belonging to the same flower. The
difference is very difficult to distinguish after the
blossom is over, and the fruit formed; but is easily
observed if the history of the fruit is traced from
the period of blossoming.

CAROLINE.

It seems to me to be very difficult to avoid error
on so complicated a subject.

MRS. B.

I cannot deny it; and I will give you another
instance of deceitful resemblance. Few things
bear a greater likeness to each other than the
Spanish chesnut (*castanea vesca*) and the horse-
chesnut (*æsculus hippocastanum*); yet the horse-
chesnut is simply a seed, while the Spanish chesnut is a fruit, consisting of two or more seeds, each
of which has its separate envelope, under the form
of a reddish-brown skin. The shell of the horse-

chesnut is a capsule produced by a single flower.
The prickly covering of the Spanish chesnut is
an involucrum, which surrounded the several
flowers. You see, therefore, that it is very difficult to decide upon the nature of the fruit without
having studied the flowers whence it derives its
origin.

Are you desirous to have another curious example of the cohesion of fruits produced by different
flowers? It is afforded in the pine-apple. This,
which you have doubtless hitherto considered as
a single fruit, is the result of the soldering of a
number of small fruits, produced by sessile flowers
aggregated on an axis, which is the stalk. These
small fruits, being soft and fleshy, unite together;
but traces of the different fruits are seen on the
surface, each forming a small protuberance: the
axis of the fruit terminates in a crown of leaves,
which surmounts the whole.

CAROLINE.

But where are the seeds?

MRS. B.

Cultivation, I have told you, tends to diminish
the quantity of seed: in the pine-apple it makes
them fail completely, so that the plant can be propagated only by the crown or by suckers. You
may see towards the centre of the pine-apple the
vacant cells in which the seeds have perished, and
in which they are lodged in the wild pine-apple,

whose fruit is less succulent and less highly flavoured.

From the pine-apple and the mulberry you
may conceive a very good idea of the fruit of the
bread-tree, which supplies the inhabitants of the
South Sea Islands with food. It may be compared
to a very large mulberry, composed of aggregated
fruits. When the seeds fail, which is the case in
the Friendly Isles, the fruit grows to a prodigious
size: when the seeds are perfected, it is in a great
measure at the expense of the fleshy part, whose
place they occupy, and the fruit is consequently
inferior both in size and flavour. This is the case
with the wild bread-tree (*Artocarpus incisa*).

EMILY.

You have said that some carpels do not open to
shed their seeds : how, then, can these sow themselves and germinate?

MRS. B.

Fruits, in this respect, may be divided into three
classes.

First. Those which do not open, and which
contain but one, or at most a very few seeds ; such
are the fruits of the gramineous family, and of
compound flowers. They are distinguished by the
name of *Pseudosperma*, which signifies false seeds ;
because, though they assume the appearance of
seeds, yet, being surrounded by their pericarp, they
are in reality fruits, and in this state they are sown
and germinate.

Secondly. Those fleshy fruits which do not open naturally ; these in the course of time become rotten, and thus disengage their seeds.

Thirdly. Fruits which are not fleshy, and which contain a number of seeds, are collectively distinguished by the name of *capsular* or *dehiscent* fruits. These open naturally and shed their seeds, which are dispersed in falling, and thus have a greater chance of germinating.

CAROLINE.

This classification of fruits is more easy to comprehend than the others.

MRS. B.

True, but it is much less important; for, instead of explaining the essence of things, it shows only the consequences. It is, however, far from being devoid of interest; but I shall enter into no further details: it is better to rest satisfied with the knowledge of a few principles, which I trust you will find no difficulty in applying to the different plants which may come under your notice. I can never sufficiently repeat, what my professor of botany has so often observed, that natural history can be learned but in a very imperfect manner in books; and that, in order to obtain a competent knowledge of objects, they must be studied in nature.

CONVERSATION XX.

ON THE SEED.

MRS. B.

BEFORE we proceed to treat of the germination of the seed, we must examine its internal structure. A seed may be considered as a germ situated at the axilla of a leaf.

CAROLINE.

Of that famous little leaf which performs so great a part in the flower and the fruit, and undergoes as many transformations as harlequin in a pantomime?

MRS. B.

No; the one I allude to is another little leaf, which adheres so closely to the germ as to form the coating of the seed itself: it is called the Spermoderm, from two Greek words, *sperma*, signifying seed, and *derma*, skin.

The spermoderm, like the pericarp, is composed of three coats.

Derived, no doubt, from the two surfaces, and the pabulum of the leaf, of which it is formed.

MRS. B.

Precisely. The external skin, called *Testa*, or *cuticle*, corresponds with the epicarp; the cellular coating, denominated *Mesosperm*, with the mesocarp; and the internal skin, called *Endopleura*, represents the endocarp.

When this leaf first shoots, it is hollow, and contains a nutritive juice called *Amnios* : the germ attached to its axilla, when fructified, begins to absorb this fluid: it takes the name of embryo; and is, in fact, a plant in miniature. In proportion as the amnios diminishes, the embryo fills out and occupies the vacant space: in the course of time it grows so large as to distend the spermoderme itself. Here is a very young bean : I slit open the spermoderme, and you see the embryo plant surrounded by the amnios.

EMILY.

But it is the miniature of a bean, not that of a plant.

MRS. B.

It is the cotyledons of the embryo plant which form the greatest part of this little bean: the radicle and plumula are enclosed within them, and are not sufficiently developed to be distinguished without the aid of a microscope. But here is a full-grown bean, in which the embryo occupies the whole interior of the spermoderm, the amnios having been all absorbed. Now, if you separate the cotyledons, you will perceive the skeleton of the plant lodged between them, and making a slight indenture in either cotyledon.

CAROLINE.

I see it perfectly; but it is not in the centre of the bean.

MRS. B.

No; it is situated at that end by which the bean was connected with the pod by a short pedicel. This spot is commonly called the eye or hilum of the seed. The pedicel conveys nourishment to the embryo plant. When the seed is ripe, this communication ceases, the pedicel withers and dries, and the seed detaches itself. This scar which you see on the testa, and which interrupts its uniform smoothness, is made by the rupture of the pedicel, and is always considered as the base of the seed; and you may still perceive the small aperture through which the nutritive juices passed into the seed.

EMILY.

But is not the embryo plant nourished by absorbing the amnios?

MRS. B.

Not wholly; for you must consider that it not only requires food for its immediate sustenance, but lays up a store of provision in its cotyledons, which is reserved for its future growth at the period of germination.

I always thought that those little threads which fastened peas and beans to the pods were merely to prevent their rolling about in the shell; but now I see that it is necessary they should have a communication with the pod, for the conveyance of nourishment.

EMILY.

What vessels in miniature these must be! I know nothing more curious than the extreme, I may almost say the invisible, minuteness of some of the organs of plants.

MRS. B.

In some seeds, the whole of the amnios is consumed by the embryo plant; in others, the absorption of this liquid is only partial: the most fluid parts pass into the embryo, while the more solid particles, being probably too bulky to traverse such minute vessels, are deposited in the interior of the seed. This substance is, at first, of the colour and consistence of the white of egg, and has thence acquired the name of albumen; but, as the seed approaches maturity, it coagulates, and adheres to the endopleur, lining it throughout with a white concrete substance, and, indeed, filling the whole of the space which is not occupied by the embryo plant. This is a resource afforded by Nature for the germination of seeds which have not a sufficient store of food in their fleshy cotyledons.

CAROLINE.

But peas and beans are so well supplied by these cotyledons, that they are in no want of such resource.

MRS. B.

Very true: the whole of the leguminous and the cruciform family, as well as several others, have no albumen. But the gramineous family, which includes all the various species of corn and grasses, are in great need of this auxiliary; for not only do they belong to the class of monocotyledons, but their single cotyledon is so small, that, although slightly fleshy, it affords but very little nourishment. But let us seek for an example on a larger scale: — you have, I dare say, eaten the white substance which lines the shell of the cocoa-nut?

CAROLINE.

Frequently: it has the consistence, and somewhat the taste, of an almond. This, then, is albumen; but what is the water that fills the cavity of the nut? — It cannot be the more fluid part of the amnios, as this, you say, is absorbed when the albumen is deposited.

MRS. B.

The seed of the cocoa-nut is very large, and

the embryo plant very small; so that the latter
cannot absorb the whole of the amnios, and it is
the residue which constitutes the water of the
cocoa-nut. Albumen, you will observe, does not,
like the cotyledons, constitute a part of the embryo plant; it is merely a deposition of food for
its use. The embryo is in general much larger
in seeds which have no albumen.

CAROLINE.

Of course, such embryos carry their store of
food about them, as a snail carries its house upon
its back: they must therefore occupy more space;
and, the whole cavity within the spermoderm
being vacant, they have more space to occupy.

MRS. B.

All that is contained within the spermoderm,
whether it consist of the embryo plant and albumen, or whether of the embryo plant alone, is
called the nucleus, kernel, or almond of the seed.

EMILY.

The amnios, then, either in its entire substance,
or a fluid secretion from it, is destined to feed the
embryo plant, while the young seed is embosomed
in the flower. The albumen and cotyledons afford
a coarser sort of food, reserved for the future nourishment of the seed when it germinates.

MRS. B.

So Nature designed it; but art converts the

greater part of this coarser sort of food into nourishment for a superior order of beings. In peas and beans it is the fleshy cotyledons that we eat; in corn it is the albumen of the seed which supplies us with bread.

CAROLINE.

But in peas and beans it is the seed itself we eat, not the cotyledons?

MRS. B.

The cotyledons form the principal part of the seed; of those, at least, which have no albumen. If, instead of eating them when young, we allowed them to ripen and germinate, the pea and the bean would separate into two parts, and assume the form of cotyledons.

EMILY.

We then rob the young plant of its destined food?

MRS. B.

No doubt we do. If corn were not reaped, the grain would fall into the ground, and, there germinating, the albumen of the seed would be expended in nourishing the young plants; but when these plants struck root, the soil would be unable to maintain a crop so thickly sown: many seeds would perish for want of food, and the rest, being but imperfectly supplied, few or none would come to perfection. Man, therefore, deserves no reproach, even from the vegetable kingdom, when he scatters the seed in such quantity only as the soil can nourish, and reserves the remainder for his own use.

EMILY.

It appears to me surprising, that the embryo plant, after having been in an active state of vegetation while the seed remained within the flower and the fruit, should become, as it were, dormant when the seed is mature, and separated from the plant; nay, should often remain so for a long period of time.

MRS. B.

The principle of life, it is true, can be preserved in some seeds a great number of years; but what that living state is, which so nearly resembles death, we cannot explain. It is time, however, for us to rouse the inactive seed from its torpid state, and examine it, when it enters into a new existence, as a separate and independent being.

EMILY.

True; we have hitherto considered only the formation of the seed, and its growth in the flower and the fruit.

MRS. B.

Let us now, then, suppose it to have attained a state of maturity, and ready, when placed under favourable circumstances, to germinate. For this purpose the seed must first be detached from the parent plant.

CAROLINE.

That is what we every day witness. The fruit, when ripe, drops from the tree; or the pericarps, when dry, burst open, and shed their seeds.

MRS. B.

Not always: some pericarps, we have observed, have no natural mode of opening; such is the nut, the amaranth, the pericarps of compound flowers, and those of gramineous plants. In the latter, the pericarp adheres so strongly to the seed, that they are confounded together, and cannot be distinguished. The seed is, in this case, inaccurately said to be naked; when ripe, it falls from the stem, inclosed in the pericarp, and, thus covered, sows itself in the ground.

CAROLINE.

Then I think that seeds of this description should be called clothed, rather than naked.

MRS. B.

They are so, in fact; but as the pericarp is of a hard dry nature, and adheres closely to the seed, it is commonly considered as forming a part of it. Thus the seeds of corn and grasses are sown enclosed in their pericarps.

EMILY.

Then the pericarp, I suppose, rots in the ground, and the seed is left at liberty to germinate?

The pericarp ultimately rots, but not until the germ has made its escape through a small aperture, which nature has provided for that purpose. That of a grain of corn is too minute to be seen with the naked eye; but you may probably have observed three openings of this description in the cocoa-nut, a seed of sufficient size for them to have attracted your attention. Through one of these the embryo escapes from its prison.

But the stem and the root cannot both shoot from the same opening, or they would both grow in the same direction?

The radicle first sprouts from the aperture with the neck situated at its base; from this vital spot the plumula shoots upwards; but the young plant remains attached to the pericarp by the neck, until it has consumed the albumen of the seed, and is able to supply itself with food from the soil.

It is thus that monocotyledons are ushered into life. The germination of dicotyledons is some-what different. The seed is not enveloped in its pericarp, and, when it begins to germinate, the spermoderm cracks and falls off; the cotyledons, commonly called the lobes of the seed, are split asunder by the stem which rises between them; but,

Plate IV.

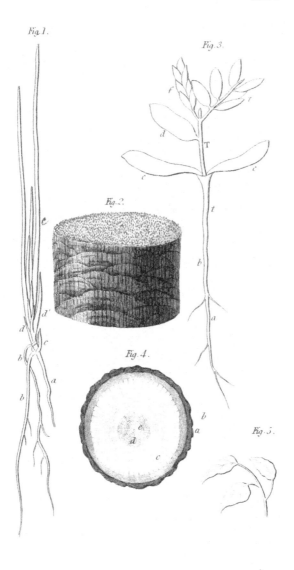

Fig. 1.

Fig. 3.

Fig. 2.

Fig. 4.

Fig. 5.

Pub.ᵈ by Longman & Cᵒ. June 22, 1829.

like a careful parent, they follow their nursling at
its entrance into life, and continue to supply it with
food until its roots are sufficiently strong to perform
that office.

The embryo plant consists, then, of three parts :
the *radicle*, or root; the *plumula*, or little stem; and
the *cotyledons*, or seminal leaves, which make their
appearance at the base of this stem.

The first and most essential circumstance requisite for germination is moisture; for a seed, in
germinating, absorbs about once and a half its
weight of water.

EMILY.

This is, no doubt, for the purpose of softening
and dissolving the hardened contents of the cotyledons, and rendering them sufficiently limpid to
pass through the minute vessels which convey them
into the embryo plant.

MRS. B.

Yes; moisture is equally necessary, whether the
germinating plant be fed by the farinaceous matter
of the cotyledons or by albumen; for seeds, when
ripe, you know, are perfectly dry, or if they contain any water, it is not in a state of liquidity, but
solid, like the water of crystallisation in mineral
salts. If seeds are deficient in moisture, they are,
on the other hand, overladen with carbon, so that
you must supply them with water, and free them
from a portion of their carbon, to enable them to
germinate. It is the great quantity of carbon

which seeds require in ripening that exhausts the soil in which they grow.

CAROLINE.

But for what purpose do they require this accumulation of carbon, since they must part with it in order to germinate?

MRS. B.

Carbon is a great antiputrescent, and is necessary to prevent the seed from rotting previous to being sown. Some seeds are, through its influence, capable of being preserved several centuries; while others, which are but scantily supplied with it, must be sown as soon as ripe. And in seeds which have not acquired a due supply of this preservative, the principle of life is extinguished before they separate from the parent-plant.

EMILY.

With a view of ascertaining whether seeds are capable of germinating, I have seen gardeners throw them into water; discard those which swam on the surface as worthless, and sow only those which sunk. They judged by the weight of the seed, I suppose, whether it contained a sufficient quantity of carbon to have preserved the vital principle.

MRS. B.

Or, rather, they know by experience that heavy

seeds are the most likely to germinate. Immersing seeds in water has also the advantage of preparing them for germination, by supplying them with the moisture of which they stand so much in need.

EMILY.

And it is, I suppose, the oxygen of the atmosphere which performs the office of relieving them from the excess of that element with which they are incumbered?

MRS. B.

Yes; it is therefore necessary that the soil should lie loosely and lightly over the seed, in order that the air should have access to it. The oxygen of the atmosphere then combines with the carbon of the seed, and carries it off in the form of carbonic acid gas. Seeds will germinate in contact with air which contains from one eighth to one third of oxygen: if the proportion be less, it will be insufficient to perform the function required; if more, the excitement will be too great, and the seed will perish from exhaustion.

EMILY.

The proportion of one fifth of oxygen, which the atmosphere contains, is, then, just the desirable medium. And heat, I conclude, is also essential to germination?

MRS. B.

To a certain degree: seeds cannot germinate during a frost, for the water must be in a liquid state: about ten degrees of Reaumur, or fifty-five of Fahrenheit, is the temperature most favourable to the germination of plants in these climates. It is, moreover, requisite that the soil should be sufficiently permeable for the slender plumula, and the tender roots, when first shooting from the seed, to penetrate it; and, on the other hand, it must be sufficiently compact to support the roots and stem when full grown. The looser the soil is, the deeper the seed should be sown, in order to afford more support.

EMILY.

And in very loose soils the air has freer access, so that there is no danger of depriving the seed of oxygen by sowing it deep. Large seeds, I suppose, require to be sown deeper than small ones?

MRS. B.

Yes; but the largest should not be buried more than six inches in the ground, in order that the air may have access to them. Small seeds require to be merely covered with earth, in order to prevent the wind from scattering them, and to shelter them, in some measure, from the light.

CAROLINE.

Is light, then, injurious to germination?

MRS. B.

The function of light, you may recollect, is to subtract oxygen from the plant, and occasion a deposition of carbon. Now, in germination, it is just the reverse which is to be effected.

When the seed, by absorption, has accumulated a sufficient quantity of moisture, it swells, bursts, the radicle shoots downwards, and the plumula rises in the opposite direction: the one becomes a root, the other a stem; and the almond of the seed is transformed into cotyledons. If any of these parts are destroyed, the plant is no doubt injured, but Nature will restore them by fresh shoots. The neck, or vital spot which forms the junction between the stem and the root, being the only part, the destruction of which proves fatal to the plant.

EMILY.

Is it known why the stem always rises, and the root descends?

MRS. B.

The roots, you must recollect, grow only at their extremities; and these, being at first of so soft a texture as to be almost liquid, naturally follow the direction of gravity and descend, unless they encounter some obstacle, such as a stone or clod of earth, so compact that they cannot penetrate it; in which case they grow out laterally, in order to avoid what they cannot overcome.

Mr. Knight performed a very curious experiment, with the view of ascertaining whether it was gravity which made the roots of a plant grow downwards. He sowed seeds in moss disposed in cavities, arranged on the circumference of a water-wheel. The cavities were open on both sides, so that the root and the stem were free to germinate at either. The wheel was then made to revolve one hundred and sixty times in a minute. The roots invariably struck in the direction diverging from the centre, like the spokes of a wheel: whence Mr. Knight was led to conclude, that, in this artificial process, the centrifugal force had replaced that of gravity.

EMILY.

That was a most ingenious contrivance. But the stem, on the contrary, grows upwards, and throughout its length.

MRS. B.

Let us suppose that it were free to grow in any direction. Since it shoots from the upper surface of the neck, it cannot grow downwards: it must, therefore, either rise vertically, or shoot out sideways. In the latter case, it will be gradually brought to a vertical direction by the same cause which makes branches tend to grow upright; that is to say, the fluids which circulate in the stem having naturally a tendency downwards, some portion, however small, will exude from the upper to the under side of the lateral stem; so that the lower, being more amply supplied with juices, will vegetate with more vigour, and grow larger. The diminutive upper side will act like the cord of a bow, and make the stem approximate towards a vertical direction; and this cause, continuing to act on the stem so long as it is not upright, will ultimately render it erect.

Let us now consider more particularly how seed should be sown, both in the fields and in gardens. In the former, the husbandman must prepare the land by ploughing, in order to render it as light as possible: the more it is pulverised, the more favourable it will be to germination. Choice must then be made of the finest grain.

CAROLINE.

Is it not considered as advantageous to change the grain, and not sow that which grew in the same soil the preceding year?

MRS. B.

It is proper, we have observed, to vary the nature of the crop; but when, in the course of cropping, grain is to be re-sown, I believe that it is perfectly immaterial whether the seed sown was grown on the same land or elsewhere.

The seasons for sowing are in spring and in autumn. It is advisable to be done early in **either** season, especially in the latter, in order that germination should take place before the frost sets in. In the spring the period must be regulated by the nature of the season and the climate. The seed may either be sown by the hand or by a drill: M. De Candolle prefers the latter, as being more exact and regular in its operation. Care must be taken not to sow too thickly. When more seed is thrown into the earth than it can nourish, part of it will perish. But this is not the only loss; for, before it perishes, it will have consumed a portion of the nourishment which otherwise would have gone to the support of the surviving crop, and which, in consequence of this subtraction, cannot attain that vigour and perfection which is natural to it when well supplied with food.

EMILY.

This is very similar to the evil effects of the excess of population, which you explained to us in your Lessons on Political Economy, when poor weakly children perish, after having languished a few years, and consumed the food which would have fallen to the share of the rest of the community, if these supernumerary children had not been born.

MRS. B.

The analogy holds perfectly good.

Lucerne has been transplanted to the distance of six inches from each other: and the plants growing larger, in consequence of their roots having a wider range for food, others have been transplanted to the distance of a foot; and others, again, as far as two feet asunder; and it was constantly found that the plants grew and flourished, in proportion as the distance between them increased.

Grains of wheat, sown very thin, have yielded from twenty to a hundred ears of corn.

EMILY.

There must, however, be a limit to this economy of seed.

MRS. B.

No doubt: land is not to be had at pleasure; but so long as the same extent of soil may be made to yield a better harvest, by sowing a less quantity of seed, it is no doubt highly advantageous.

The only exception to this rule is when you aim at producing long and slender stems. This is the case with hemp and flax. Comparatively little value is set upon the seed: the stems, for making linen, are the essential produce. Those seeds must therefore be sown very thick, in order that the stems may grow long and upright, and no space be allowed them to branch out.

The Italian corn, with the straw of which hats are made, is sown very thick, with the same intention, and cultivated on a barren rocky soil, in order that a deficiency of nourishment may give the straw that morbid delicacy and slender form which render the Leghorn hats so fine.

Let us now turn our attention to garden culture. When seeds are sown from foreign parts,
you may form some judgment of the degree of
temperature and nature of the soil which they
require, by the latitude and elevation of the spot
whence they came. Seeds from tropical climates
should generally be sown in hotbeds, having stone
or wooden frames: the latter, being a worse con-
ductor of heat, preserve plants from the cold
better than the former. Experience teaches us
that hotbeds are preferable to hothouses, both for
the germination of the seed, and the growth of
very young plants; and small hothouses are preferable to large ones (though of an equal temperature), so long as the plants have sufficient room
to grow. Of course, they must not be cramped and
stinted for space; for large plants require extensive accommodation: but the reason why a confined space is advantageous to small plants has
not hitherto been ascertained. The heat generated
by the fermentation of manure, is also more favourable to germination than the heat of a stove.

CAROLINE.

That, I think, is easily accounted for. The heat
of a stove is of a drying nature, whilst that of the
fermentation of manure is always accompanied by
moisture, which will accelerate the swelling of the
seed and bursting of its coats. And why should
not this be the reason, that a hotbed is preferable
to a hothouse, for the purpose of raising plants

from seed ? For the one is heated by fermentation, the other by a stove.

The pots in which they are sown are frequently placed in beds of manure in a hothouse. Besides, the same argument holds good with regard to greenhouses: the smaller the house, the better it is calculated for the culture of small plants. It has been suggested, that small plants being always placed in the front and lowest rows of the greenhouse, and hot air having a tendency to rise, they occupy the coldest strata of air.

In the preparation of hotbeds, the manure of horses is preferable to that of cows, as it more readily decomposes, and enters into combination with vegetables. That of pigeons possesses this quality in such a remarkable degree, that it is dangerous to germination, by precipitating it too much.

The seeds, however, are, I believe, never sown in manure itself, but in pots of earth which are sunk in it; the plant, therefore, benefits only by the heat and the evaporable particles.

That is true. Little or no manure should be mixed with the earth contained in the pots, in which germination takes place; for the seed, at

that period, far from being in want of food, requires to get rid of a surplus of carbon.

EMILY.

But when the germination of the seed is completed, and the roots shoot out in search of food,
some provision should be made for them.

MRS. B.

If the earth contained in the pots consists of rich
garden-mould, it will afford a sufficiency. When
land is manured for grain, the seed derives no
advantage from it: the embryo plant is nourished by the albumen; and it is not till the roots
have acquired some consistence and vigour that
they begin to supply the plant with food from
the soil.

You must observe that it is necessary to cover
the hole, which is made at the bottom of all garden-
pots, with a small piece of tile; and it is proper,
also, to place a second piece, of larger dimensions,
over the first, in order effectually to prevent the
too rapid filtration of water.   Care must be taken
to keep the earth light and loose over seeds which
are germinating; for if the soil be calcareous in
drying, after being watered, a crust frequently
forms on the surface, through which the slender
stem cannot penetrate, and the young plant is thus
literally buried alive. The surface of the mould
must, therefore, be kept scratched or raked, to

prevent this crust from forming; or to pulverise it, if the evil has already taken place.

Shallow wooden boxes are frequently used, instead of pots, for the purpose of sowing seeds: they have the advantage of affording them more space.

When plants have so far increased in size as to require transplantation, they should not be pulled up by the roots, but the whole clod of earth be carefully shaken out of the pot at once, and then gently divided into parts, so as to run no risk of wounding the fibrous extremities of the roots in separating them from the earth which surrounds them.

EMILY.

I have observed that, in transplanting them, the gardener uses a pointed instrument to make a hole in the ground, and afford room for the roots of the young plant.

MRS. B.

This is exactly the reverse of ploughing: it makes an opening to receive the young plant, no doubt, but it is at the expense of the contiguous soil, which is rendered proportionally more compact. It is true that the young vegetable, at the period of transplantation, has acquired some vigour; and a light soil is not so essential to it as during germination. It is preferable, however, to transplant in furrows, when the earth is turned up with a spade or a hoe, and the roots afterwards covered by raking the earth over them.

F 5

But we are deviating from our subject. Now that we have fairly traced the seed through the process of germination, we should conclude our conversation, and reserve what remarks I have to make on planting to some future day.

CONVERSATION XXI.

ON THE CLASSIFICATION OF PLANTS.

EMILY.

You have often talked to us of plants belonging
to different families, my dear Mrs. B., but you
have never explained the exact meaning of the
word family in the vegetable kingdom; and I
wish you would also teach us, how to find out, to
which family a plant belongs.

MRS. B.

Your question is much more complicated than
you are aware of; and I know not how to give you
a satisfactory answer without explaining the whole
theory of classification, which will be long, and
sometimes, perhaps, you may find it tedious.

EMILY.

I have no fear of undertaking it; for I am conscious it is necessary, and that, without such information, it would be difficult to remember many
things that you have taught us.

MRS. B.

You are quite right: the number of natural
beings is so immense, that without some mode of
classification it would be impossible to form a
correct idea of them.   Would you believe that
there are no less than sixty thousand species of
plants already known, and that this number is
increasing every day?

CAROLINE.

Oh, Mrs. B., what a host!

MRS. B.

You have named it very justly.   Well, then, to
carry on your comparison of a host, do you conceive that any general, in reviewing sixty thousand
soldiers, would be capable of recollecting every
individual?

CAROLINE.

Certainly not; but as each regiment has its
uniform, and each company its number, he could
easily discover to which any soldier belonged.

MRS. B.

This is exactly the object a botanist has in view
in classification.   He endeavours to find out to
which regiment and to which company each individual plant belongs; but this is much more
difficult than with an army, where the general has
himself chosen the uniforms, and arranged the
companies so as to make the distinctions most con-

spicuous.   Nature has also, it is true, her distinguishing characters, but they are often placed in
organs which are least exposed to view; and botanists do not always agree on the characteristic
features of plants.

CAROLINE.

Well; but tell us at least on what points all
botanists do agree?

MRS. B.

Willingly.   You see yonder in the garden a
bed of carrots: every individual plant of which
belongs to the same species.

CAROLINE.

Of course; for I know that the gardener collected the seed sown in that bed from one individual plant last year, so that the carrots must all
be descended from the same parent.

MRS. B.

Now, then, if you extend your idea to all
the carrots in the world, which do not differ from
the carrots in this bed, more than these differ
from each other, you will understand that they
may originally have been derived from the same
plant.

EMILY.

Certainly; and that is what we have said con-
stituted a species.

MRS. B.

Well, then, if this idea is clear to you, let us proceed a step further, and, considering each species as a unity, compare them with one another. Can you recollect any instance of different species bearing a striking resemblance to each other?

CAROLINE.

Oh, yes; there is the white rose, the yellow rose, the China rose, and many others, which are very much alike; and yet the gardener assures me that the seed of the one will never produce the other.

EMILY.

Red clover, white clover, and yellow clover, bear also a similar resemblance to each other, though they are all of different species.

MRS. B.

Your examples are well chosen; for the most ignorant person could understand them. Species bearing this analogy are classed together under the name of *Genera*. The resemblance of the different species composing some of the genera is so striking, that their affinity cannot be mistaken; while in others it is less marked, and requires some study to be recognised.

CAROLINE.

Like different children of the same family, some

of whom are so much alike that you see at once they are brothers and sisters; while, in others, the relationship cannot be traced in their features.

MRS. B.

The word *genus* is a Latin word, signifying family: your comparison, therefore, cannot but be accurate; and, by following it up, I think you may acquire an idea of the whole system of nomenclature. You may easily conceive, that if there were a separate name for each of the sixty thousand species of plants, no memory could retain them; nor could these names point out the resemblances or differences which exist between the several species.

EMILY.

Of course; just as if every individual of a country had a different surname: it would be impossible to know, among which of them, any relationship existed.

MRS. B.

I see that you have nearly made out the simple art of nomenclature. Each genus has a substantive name; as rose, carrot, clover: to which is added the epithet of white, black, large, small, to designate each species; so that, instead of sixty thousand names, five thousand only are sufficient: to which are added a certain number of epithets in common use, and understood by every one.

EMILY.

There is, then, really a great analogy between

the nomenclature of plants and that of mankind, for the names of the genera correspond with our family names, and those of the species with our Christian names.

MRS. B.

Very true. It is to the celebrated Linnæus that we are indebted for this simple and clear mode of nomenclature; and it is one of the circumstances which has contributed most to facilitate the study of botany. Since, then, the basis of nomenclature rests on the idea of genera, you may judge how important it is that this idea should be clearly understood, and that plants not possessing the requisite analogy should not be placed in the same genus.

EMILY.

It would be easy to ascertain this analogy with regard to the species, by sowing their seeds; but I know not what test there is to verify the genus of a plant?

MRS. B.

No wonder you should be at a loss, for it is a question which has embarrassed the most celebrated botanists. There is but one means of forming a clear idea of genera: it is by placing its distinguishing characters in those organs of the plant, which, in the common course of Nature, are least liable to variation. Now, it has been observed, that stems and leaves more frequently vary in their structure than flowers and fruit: botanists have, therefore, agreed to place the characters of the genera in the organs of fructification.

EMILY.

But are not flowers of the same species sometimes blue, and sometimes white; sometimes large, and at others, small? How, then, can you establish the character of the genera, on circumstances which cannot even serve to distinguish the species?

MRS. B.

The difference or resemblance of organs, must not all be considered as of equal importance, in deciding the genus of the plant. All such as relate to colour, flavour, smell, consistence, and the absolute size of the organs, must, in classification, be set aside; and those only which are connected with the symmetry of the flower or fruit, the number of parts, their shape, and relative size, are to be attended to.

CAROLINE.

Why do you consider the relative size of organs of more importance than their absolute size?

EMILY.

That appears to me quite clear. If a plant grows in a rich soil, all its parts will be larger than a similar plant growing in a meagre soil; yet the plants will be of the same nature. But if the stamens of a plant be twice as long as its petals, it will certainly be of a different nature from a plant of which the stamens are shorter than the petals; and, were the two plants well or ill fed, their proportion or relative size would not be altered.

CAROLINE.

You say, Mrs. B., that the number of organs of a plant is admitted as one of the characters of genera; but I think, I remember having seen, on the same syringa shrub, flowers, some of which had four, and others five petals.

MRS. B.

Your observation is correct; and, in order to obviate this difficulty, you must be guided by the same rule, in regard to number, which Emily has just pointed out with regard to size. Thus, for example, the pink has twice as many stamens as petals : this character never varies, unless in some peculiar cases of monstrosity, which derange the whole economy of the plant.

With regard to the absolute number of organs, that is to say, whether a plant has four or five petals, eight or ten stamens, is a circumstance attended with much more uncertainty ; it may, however, be used as a character in classification, provided it be done with caution, and in cases only, in which experience has shown, that the variations are inconsiderable. Botanists know, for instance, that the greater the number of organs, the more liable that number is to vary.   This is a law which prevails throughout all Nature.

It is evident that the number of our fingers varies less than that of our teeth, and the latter less than that of our hair.

These general rules will, I hope, make you understand how botanists have been enabled, gradually to draw a more accurate line of demarcation in circumscribing the genera of plants, and determining their characters.   This constitutes one of the most essential branches of botany; for it is certain, that the better the structure of plants is known, the more easily new plants are discovered, because we are enabled to ascertain their characters, and, consequently, to class them with greater precision ; and thus the art is gradually brought to some degree of perfection.

But since the nomenclature is founded upon genera, if you change the genera you must change also the names of plants, which is very perplexing for beginners.

That is true; but it would be a source of still greater perplexity to allow a plant to remain in a genus, in which, through ignorance, it had been erroneously placed, and of which it had not the characters, for you would never be able to recog-

nise it. Let us suppose, for instance, that the pear-
tree had been placed in a genus, the distinguishing
character of which, is to have six petals : as it has
only five, you would never think of searching for
it in a genus of six petals. Would you, then, dis-
approve of its being transferred to the genus whose
character is to have five petals ?

EMILY.

No, certainly; but such errors seem to me to
be impossible.

MRS. B.

Not so much so as you imagine, owing to the
number of the organs frequently differing in the
same plants, like the petals of your syringa. Besides, foreign plants are often described in a hurried
manner by incompetent botanists, and are sometimes brought to Europe in such an imperfect state
of preservation as to afford very bad specimens.
You see, therefore, that in this case, as well as in
many others, truth must be preferred to simple
convenience.

CAROLINE.

But, my dear Mrs. B., you have not yet said one
word of the families of plants, which you had promised to explain to us.

MRS. B.

A little patience, my dear. I cannot tell you
every thing at once; and I like to begin by the
beginning. I have explained to you how all indi-

vidual plánts, resembling each other, form a species; and how, by following up the same idea, and uniting in your mind all the species whose flowers and fruits are nearly similar, you form a genus. When you have accomplished this, proceed a step further: consider the genus as a unity; and we will then endeavour to make out how the five thousand genera of the vegetable kingdom can be so arranged as to be most easily recognised.   But first tell me what is your particular aim in studying botany?

CAROLINE.

I wish to be able to find out the name of any plant I may chance to meet with.

EMILY.

I confess, for my part, that I am not very anxious about the name; but I should like to understand the structure of any plant I see, and how far it coincides with or differs from other plants.

MRS. B.

You have each answered as I expected.   Caroline, who is the youngest, and the most volatile, is satisfied with enquiring after names. Emily, who is older, and more considerate, seeks after things. Well, my dears, the learned world have done like you.   In the infancy of botany, they sought after names: when further advanced, they aimed at learning the essence of things.   I shall follow their steps, and first explain to Caroline, by mere no-

menclature, how to find out the name of a plant;
and then teach Emily, what has been called, the
natural method.   But I think you must be tired to-
day : we will, therefore, defer it till our next interview.   In the mean time, reflect upon the subject;
and let me know, when we meet, which you think
may be the best means of attaining the object you
each have in view.

CONVERSATION XXII.

ON ARTIFICIAL SYSTEMS OF CLASSIFICATION OF PLANTS.

MRS. B.

I HAVE promised to explain to you to-day, Caroline, how to discover the name of a plant. But why should you not enquire of some botanist the names of the plants you may wish to know, and then fix them by rote in your memory?

CAROLINE.

Oh, Mrs. B., you think me still more childish than I really am. In the first place, is it probable that I should always meet with a person capable of telling me the name of a plant? And, then, I should be liable to be imposed upon and misinformed. No; what I desire is to be enabled to find out the names of plants myself — with the assistance of books, I mean.

MRS. B.

It would be necessary for this purpose to have

a dictionary arranged the very reverse of those commonly used. Instead of giving the word, and then the definition, as dictionaries usually do, you must begin by learning the characters of plants, and then come to the name.

CAROLINE.

But a dictionary of this description would be but of little use to me; for, as I am completely ignorant of the characters of most plants, I should not know where to seek for them in the dictionary. Indeed, my principal reason for wishing to know their names is to be able afterwards to learn their history, and become acquainted with their uses and properties.

MRS. B.

I see that Caroline is coming round to your opinion, Emily; she proceeds from names to things. This shows me that you have reflected upon the subject, since our last interview. Various modes have been resorted to, with a view of composing the peculiar species of dictionary to which I have alluded; and, in order to give you some idea of them, tell me, have you ever played at a game called *Four-and-twenty Questions?*

CAROLINE.

Oh, yes; I am quite an adept at it. One of the party thinks of a word or a thing, and the others try to discover it, in a series of twenty-four questions.

MRS. B.

In this game, the art consists in so skilfully pointing the questions, that each reply shall contract the field of enquiry, till it is at length brought within so small a compass, that the object thought of is attained.

EMILY.

It is like the Indian mode of entrapping elephants. They are surrounded by horsemen, who at first enclose a large space of ground, but, gradually narrowing the circle, drive them gently towards the centre; where they have no resource but to run into the trap, which is there open to them.

MRS. B.

Well, botanical nomenclature is very much the same thing. Indeed, the analytical method — the simplest hitherto used — is exactly similar to the game of Four-and-twenty Questions. It was first introduced by M. De Lamarck, and forms the basis of a work he published, with M. De Candolle, called the *Flore Françoise*.

Here is a plant in blossom in this flowerpot: I know that it grows wild in the south of France. I shall ask you a few questions relative to it, according to the method followed in the *Flore Françoise*, and you will see that you will soon discover the name of the plant yourself.

CAROLINE.

Oh, that will be very amusing.

MRS. B.

First, then, tell me whether the flowers are visible to the naked eye, or whether they require to be seen through a microscope?

CAROLINE.

The plant is quite covered with them, and they are perfectly visible; for you may see them from some distance.

MRS. B.

Are the flowers united by an involucrum or not?

CAROLINE.

No; they have no involucrum.

MRS. B.

Have the flowers both pistils and stamens, or only one of these organs?

CAROLINE.

They have both.

MRS. B.

Have the flowers both a corolla and a calyx?

CAROLINE.

Yes; they have both.

MRS. B.

Is the corolla of one entire piece, or composed of several parts?

CAROLINE.

It consists of several parts.

MRS. B.

In this case, I must refer to No. 211. of the *Flore Françoise*; and pray tell me, does the ovary adhere to the calyx or not?

CAROLINE.

It adheres to the calyx, and is consequently situated below the petals.

MRS. B.

Are there several ovaries?

CAROLINE.

No; only one.

MRS. B.

Is the flower regular or irregular?

CAROLINE.

Regular.

MRS. B.

Are there more or less than twenty stamens?

CAROLINE.

There are, certainly, more than ten; and I should think about twenty.

MRS. B.

Is the calyx divided into more than two lobes?

G 2

CAROLINE.

Yes; into five.

MRS. B.

Are the leaves opposite or alternate?

CAROLINE.

They are opposite, and placed very regularly on the stem.

MRS. B.

Is your plant herbaceous, or is it a shrub?

CAROLINE.

A shrub.

MRS. B.

Is the fruit dry or fleshy?

CAROLINE.

Let me see if I can find any: there is very little fruit; but it appears to me fleshy.

MRS. B.

Is that part of the calyx which crowns the fruit membranaceous or coriaceous?

CAROLINE.

It is membranaceous.

MRS. B.

Well, then, your plant must be a myrtle. Were there several species of myrtle growing in France, by continuing a similar series of questions, I should soon discover the species; but as there is only one (which is the common myrtle), I am already fully answered.

CAROLINE.

What a simple and ingenious method! You would easily have won the game of Four-and-twenty Questions; for you have asked only thirteen. This mode of analysis must, I suppose, be in very general use?

MRS. B.

It has been much used in France, but very little in England; nor do I know any English work on botany arranged according to this method.

CAROLINE.

I am surprised at that. Is there any thing to be objected to it?

MRS. B.

It has been thought long and tedious; and, to one no longer a novice in botany, it is tiresome to have always to recommence and follow up the same routine of questions; besides, the slightest inattention, or the least mistake in the printed numbers, is sufficient to put you quite out. Then, after all, when you have succeeded in discovering the name of the plant, it is not very easy to remember the characters which enabled you to find it out. Notwithstanding these objections, which certainly have their weight, I cannot but think this method the best for beginners.

CAROLINE.

Still, if it is not used in England, it would not be of much use to us. Pray, what mode of nomenclature is adopted by English botanists?

MRS. B.

That which we owe to Linnæus, which I have here, and will show you.

EMILY.

Pray do; for the name of Linnæus is so celebrated, that any method invented by him cannot but be interesting.

MRS. B.

You are right. A method derived from a man so eminent in science, and which has been adopted by so many other scientific men, well deserves our attention. Here is a table of the different classes into which Linnæus divided the vegetable kingdom.

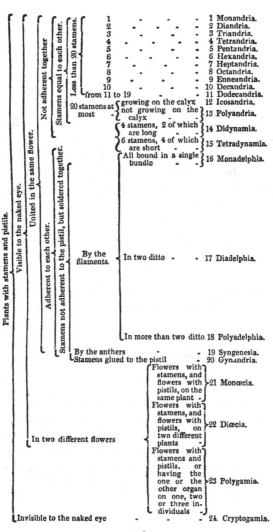

	1 Monandria.
	2 Diandria.
	3 Triandria.
	4 Tetrandria.
	5 Pentandria.
	6 Hexandria.
	7 Heptandria.
	8 Octandria.
	9 Enneandria.
	10 Decandria.
from 11 to 19	11 Dodecandria.
20 stamens at most { growing on the calyx	12 Icosandria.
not growing on the calyx	13 Polyandria.
{ 4 stamens, 2 of which are long	14 Didynamia.
6 stamens, 4 of which are short	15 Tetradynamia.
{ All bound in a single bundle	16 Monadelphia.
By the filaments. { In two ditto	17 Diadelphia.
In more than two ditto	18 Polyadelphia.
By the anthers	19 Syngenesia.
Stamens glued to the pistil	20 Gynandria.
Flowers with stamens, and flowers with pistils, on the same plant	21 Monœcia.
In two different flowers { Flowers with stamens, and flowers with pistils, on two different plants	22 Diœcia.
Flowers with stamens and pistils, or having the one or the other organ on one, two or three individuals	23 Polygamia.
Invisible to the naked eye	24. Cryptogamia.

CAROLINE.

It appears very complicated !

MRS. B.

It is less so than you imagine. This system of
Linnæus forms precisely an analytical table. Return to your myrtle, and let us follow it. — Stamens visible to the naked eye, in the same flower
as the pistils, but not united to them; all of the
same size; about twenty in number; growing from
the calyx: the myrtle belongs to the class Icosandria of Linnæus.

CAROLINE.

And then what follows ?

MRS. B.

These classes are divided into orders, which
depend principally on the number of styles.

CAROLINE.

The myrtle has only one style.

MRS. B.

Therefore it belongs to the order Icosandria
Monogynia. Here is an edition of Linnæus, published by Willdenow. You see that the above
order contains twenty-two genera: you must therefore examine the characters of each genus, and you
will soon find out those which correspond with the
myrtle.

EMILY.

This method, certainly, must be much more complicated than the other.

MRS. B.

It may appear so to a beginner ; but you would soon become used to it : and I assure you that it is a very convenient mode of discovering the names of plants; at least, of those which do not belong to a class containing a very great number of genera; such, for instance, as those of *Pentandria* and *Syngenesia*, which contain from four to five hundred each. But, even in these cases, practice and habit soon render familiar a system which at first sight appears perplexing and difficult.

EMILY.

How often have I made that observation ! When I first began learning the piano-forte, I shall never forget the difficulty I had to distinguish the different notes; whilst now I play without ever thinking of them. But, Mrs. B., the system of Linnæus rests almost entirely on the number of the different organs of plants ; and we have observed more than once, that number is a character which can be little relied on in botany. Yesterday we mentioned the irregular number of petals of the Syringa; and this morning I observed a plant of rue which bore in the same cluster flowers, some of which had eight, and others ten stamens. Does it belong to the eighth or the tenth class ? I am sure I deserve

G 5

to know, were it only to make up for the disagreeable smell I had to encounter in examining it.

MRS. B.

In cases of this sort, botanists have agreed to
class the plant according to the first flower which
blows in the cluster; and, following this rule, the
rue belongs to the tenth class. But what is still
more embarrassing is, that the number of stamens
often vary in an irregular manner. Thus you may
meet with tulips, and several other plants, bearing
indifferently five, six, seven, or eight stamens.
In this case you must class them according to the
number most usually found.

Several other sources of error occur in the system of Linnæus. For instance, there are many
plants in which the inequality of the stamens, and
their adherence, is so difficult to distinguish, that
it is not easy to know to what class they should be
referred. There are also *Diœcious* or *Monœcious*
plants, which are become so by mere accident; and
which you would seek for in vain under the head
Diœcia or *Monœcia.*

The class *Polygamia* is also one of those which
can scarcely be made out by a beginner. Notwithstanding the imperfections of the Linnæan
system, it is one of the most convenient; and, even
were it less so, it would be necessary, in order to
make any proficiency in botany, to be well acquainted with it, as it is that which is most generally used in botanical works.

EMILY.

But, in small flowers, does not the extreme minuteness of the organ which it is necessary to investigate, render the system of Linnæus liable to error ?

MRS. B.

True; this is one of the sources of error which even the genius of Linnæus could not overcome.

EMILY.

When two plants have each an organ perfectly similar, is it a necessary consequence that they should resemble each other in other respects?

MRS. B.

Sometimes, but not always. Thus, in this system of Linnæus, the class Tetradynamia contains plants all having a certain natural resemblance; the same may be said of the class Syngenesia; but in the other classes they frequently differ considerably.

EMILY.

This is not very logical, I think !

MRS. B.

It is true that it would not be so, had Linnæus meant that his system should point out the real differences of plants. But his intention was merely that it should answer the purpose of a dictionary, by means of which it would be easy to discover the names of plants; and, in this point of view, it may

G 6

fairly be affirmed, that the greater the number of points of difference existing between plants, belonging to the same class, the easier it is to discover the name of each.

The followers of Linnæus have unfortunately not always well understood his intention of thus supplying botanists with a mere nomenclature, or, as it has been called, an *Artificial System ;* and the habit of studying the pistils and stamens has made them attach too great importance to these organs of fructification, while they have neglected the fruit and the seed. They have also, in general paid too much attention to the number of the organs, and not sufficiently considered their relation to each other ; so that, if they have rendered it an easy thing to discover the name of a plant, they have not advanced the science of botany so much as they would have done had they directed their researches more to the general properties of plants.

EMILY.

The system of Linnæus might, I think, be compared to a dictionary, which, though you learnt all the words it contained by heart, you would still be ignorant of the language, unless you added to it a knowledge of the grammar which teaches the value of the relative terms.

MRS. B.

Well; that arrangement, which is called the *Natural System,* teaches you the grammar. But, in reply to your observation, I might retort that a grammar alone is not sufficient to make you acquainted with a language — a dictionary is also necessary, in order to look out for the words; you must not therefore undervalue botanical dictionaries, which facilitate the study of nomenclature: yet always bear in mind that they teach only the names of plants; and that, if you wish at the same time to acquire a general knowledge of their structure, you must study the *natural system*.

EMILY.

That is what I am most anxious to do: pray give us some general idea of it.

MRS. B.

With pleasure; but not to-day: we shall examine it when next we meet. In the mean time reflect upon the subject, and endeavour, of your own accord, to discover some mode of classing plants which would most easily show their true analogy.

CAROLINE.

I will think of it; but it appears to me a very difficult task.

CONVERSATION XXIII.

ON THE NATURAL SYSTEM OF CLASSIFICATION.

MRS. B.

WELL, my dear, have you been able to discover any mode of classing plants, according to the analogy they bear to one another?

CAROLINE.

I have endeavoured to class the plants in our garden according to this method. I began by comparing them all together, and then divided them into groups, according as they more or less resembled each other.

MRS. B.

I should have no fault to find with your mode of proceeding, if the whole vegetable kingdom contained, like your garden, only a small number of plants. This mode was, in fact, the first employed, and was called the *Méthode de Tatonnement.* Even now it is occasionally employed by botanists, as a guide in their researches. But you will easily understand, that, independently of the impossibility of being put in practice, since the number of plants known has so much increased, it has also the great defect of depending merely upon opinion, and affording no certainty of the reality of the resemblance assigned to different plants. It is much the same as with likenesses of different persons : how often people vary in opinion in regard to such resemblances !

CAROLINE.

So much so, that, while many people declare that I am the very picture of my father, others see no resemblance whatever between us.

MRS. B.

The same diversity of opinion would take place in natural history, had not botanists laid down certain precise rules for judging of the external characters of plants ; and, Emily, have you no new mode of classification to suggest ?

EMILY.

If I tell you the one that has occurred to me, I fear you will think me very presumptuous.

MRS. B.

By no means, my dear: on the contrary, nothing is more gratifying to me, than to see that you are capable of reflection, whatever may be the object.

We examined yesterday the system of classification founded upon a single organ : well, if I had to class the vegetable kingdom, and was well acquainted with the structure of plants, I should make as many different arrangements as there are different organs.   The first, for instance, would be made after the roots, the next after the stems, the following after the leaves, another relating to their position, another according to the number of the organs, and so on.   I should thus form, perhaps, a hundred different systems.   Now, I suppose that the plants which were placed together in ninety-nine of these systems would bear the strongest possible resemblance to each other; in ninety-eight, a little less so: in a word, that the resemblance between plants could be ascertained by the number of systems common to each.

Your idea is very ingenious, my dear Emily; and, though I do not agree with you in opinion, you may boast of having suggested the same theory as a very distinguished French botanist, M. Adamson. He called it, " Method of General Comparison," and bestowed much labour upon it; but, as he lived so long ago as the year 1760, he was far from being acquainted with all the different characters of plants, which rendered his system very incomplete. Even in our time, though great progress has been made in botany, new characters and properties are frequently discovered in plants; consequently, a classification of this description would still be incomplete.

EMILY.

I own that objection did not occur to me, because I thought that the degree of precision in any mode of classification depended upon the state of the science at the time it was made.

MRS. B.

There is another objection, of still greater weight. According to your system, you count the number of organs that resemble each other in different plants, but you make no estimate of the relative importance of each; yet you must consider that all the organs are far from being of equal importance. Plants which resemble each other in a few of their principal organs, have more real analogy than those which are similar in a great many minor points. You may easily conceive that plants whose seeds are alike, resemble each other infinitely more than those which shoot out thorns of the same nature. For the seed may be considered as the miniature of the plant, from the developement of which all the growth of it must arise; while the thorn is a mere accidental degeneration, which may or may not take place.

EMILY.

This objection is certainly of great weight, and I am afraid that I must abandon my system; but is it possible to appreciate the relative importance of the different organs?

MRS. B.

To a certain extent, at least. The method of classification, grounded on this principle, is called " Method of Subordination of Characters." It was first suggested by Bernard Jussieu.

EMILY.

How much I should like to read his work on this subject!

MRS. B.

He never published any thing: like Socrates, he taught in conversation. His nephew, M. Antoine Laurent De Jussieu, who is still living, published, in 1789, the results of his uncle's theory. In 1823, M. De Candolle published a work, in which the principles of this mode of classification are fully developed: they are probably the same as those of M. De Jussieu, since his deductions from them are similar to those of his predecessor.

CAROLINE.

Then M. De Candolle must be the Plato of this modern Socrates?

MRS. B.

Or even something more; for Plato, I believe, wrote only what Socrates had taught; but M. De Candolle brought to light those principles from which M. Bernard Jussieu had drawn his deductions.

EMILY.

It is a curious thing in science, for the founder of a new school not to publish his opinions. But, pray, what are these principles? They must, I think, be very difficult.

MRS. B.

We have agreed that all the organs of plants are not of equal importance: now there are three modes of ascertaining the degree of consequence of each. The first is its utility: this is the most general, and would be sufficient of itself, were we so perfectly acquainted with vegetable physiology, as to judge of the importance of an organ, by its degree of utility in the economy of a plant. Thus we may safely conclude, that an organ essential to the life of a plant, is of a higher order than one of which it can be deprived, without sustaining any material injury.

EMILY.

But are the functions of different organs not sufficiently known, to enable us to judge whether they are more or less essential to a plant?

MRS. B.

Not always; but in a doubtful case, there are

two other rules by which you may be guided. Generally speaking, the greater the number of plants in which the same organ can be found, the greater is the degree of importance that ought to be attached to them. For instance, the calyx may be pronounced to be an organ of much greater consequence than the involucrum, because a much greater number of plants have a calyx than an involucrum.

EMILY.

I understand that perfectly.

MRS. B.

There is besides a third criterion. There are certain organs which exist, or which are wanting, in all the plants of the same family; and others that are only occasionally found in plants, otherwise very similar to each other. The first of these organs are naturally of much greater importance than the latter, as they appear to be indispensable to the system of organisation of these plants. Were you to ask me, for example, whether a stipula or a thorn were of greatest importance, I should not hesitate to say the stipula, for the reason I have just assigned.

EMILY.

When you judge of the importance of an organ by its degree of utility in the economy of a plant, how can you compare organs adapted to functions of a completely different nature? In the animal frame, it would be difficult to determine whether the lungs were a more useful organ than the stomach, the eye than the hand — does not the same difficulty occur in the vegetable structure?

MRS. B.

Your observation is perfectly just; and, in fact, botanists can only, with any degree of certainty, compare such organs as are adapted to the same class of functions. For instance, you will readily admit that the brain is of higher rank than any single nerve, and the heart superior to any other blood-vessel: but if you enquired whether the heart or the brain were of greatest importance, it would be quite out of my power to answer you. If you will promise not to laugh at me, I will venture upon a very trivial comparison.

CAROLINE.

Pray, let us hear it, Mrs. B.; I am so fond of comparisons — indeed, I often understand them better than arguments.

MRS. B.

You are, no doubt, aware that a captain is of higher rank in the army than a lieutenant, and a colonel than a captain; you know, also, that the governor of a province is of more elevated dignity than the mayor of a small town: but, pray, how would you answer, if I asked you whether a captain or a mayor ranked highest? You might say, in some particular cases, the one takes precedence of the other; but that would depend entirely upon arbitrary decision, and not on the nature of their functions, which will not admit of comparison. Now, if you apply this simile to vegetable physiology in which there are two great classes of functions; one of which belongs to the re-production, and the other to the nutrition, of plants; you will understand that those organs alone admit of comparison which belong to the same class.

EMILY.

Pray, give us some example of this?

MRS. B.

In re-production, for instance, the organ of most importance is the embryo; next to that the stamens and the pistils, which, taken collectively, are no less indispensable than the embryo, for without them it cannot receive life. Then follow the integuments which protect the embryo; and next those which guard the pistils and the stamens; after these come the accessary organs, such as the nectary. You see that I have already given five different degrees of-importance to the organs of re-production.

EMILY.

But must you not also arrange, in a similar order of gradation, the different points of view under which a plant may be considered?

MRS. B.

No doubt; and, for this purpose, you must be guided by the rules we used in forming the different genera. The most important of which consists in carefully observing the symmetrical position of the organs in different plants. The natural method of classification consists in studying the details of the symmetry of the organs, in the same manner as mineralogy is founded, on the regular symmetrical laws of crystallisation.

EMILY.

All this appears to me very ingenious in theory, but difficult in practice. Supposing that I were capable of classing the vegetable kingdom according to this order of different organs; what proof should I have that I was following the right method?

MRS. B.

You might afterwards class the vegetable kingdom according to the organs of nutrition, and you could then compare the two arrangements. Now, if, in following two methods, each founded upon a set of different organs, the same plants are to be met with in the same class, is it not infinitely probable that the mode of classification you have adopted is the true one?—the image of what really takes place in Nature? Thus, the natural order of botany is that in which you obtain the same result, whether the vegetable kingdom be classed according to the organs of re-production, or to those of nutrition. More importance is usually attached to the organs of re-production, as being the most numerous and the most varied; the classification is, therefore, first made with reference to them, and afterwards with reference to the organs of nutrition: the latter of which serves to verify the former.

EMILY.

It is like making a proof in arithmetic: but is not this very difficult to reduce to practice?

MRS. B.

I do not deny that it is sometimes attended with difficulty. In botany, as in every other science, no progress can be made without labour and perseverance: — much yet remains to be done, but it is gratifying to have a great end in view: it elevates the mind, and renders the details of a science interesting. The difficulties that occur in classification arise, either from our not yet knowing all the plants that exist, or from our limited faculties often preventing our acquiring a competent knowledge, of the nature and internal structure of their organs. Time may overcome the former of these difficulties; but the latter will probably never be completely conquered. Sometimes, for instance, the organs of plants which ought to be symmetrical, are not all developed; at others, they are joined together so that their number cannot be distinguished : — this we have called soldering. Sometimes they assume unusual form and dimension : this is called degeneration of the organs. These three causes, considered either collectively or separately, often deceive botanists in regard to the real nature of the vegetable organs, but, by dint of observation, the truth is gradually brought to light.

EMILY.

We should, no doubt, be incapable of understanding in detail the results of the principles you have explained to us; but cannot you give us some slight idea of them ?

MRS. B.

I will make the attempt, at least. You may recollect learning the other day, that genera were composed of those species most nearly resembling each other. Now, by means of the principles I have just laid down, it was soon discovered, that in a certain number of genera the organs of reproduction were very analogous : it was then ascertained, that the organs of nutrition of these same genera, also bore a striking resemblance to each other. These genera were then united, as it were, in a group, and denominated a *family*. Thus the five thousand genera form about two hundred and fifty families.

CAROLINE.

Families of plants, then, are nothing more than

a numerous collection of genera resembling each other. But, then, *genera*, in the sense in which it is here taken, means not *families*, as it ought to do, from its Greek origin, but merely branches of families : is not this liable to create confusion?

MRS. B.

I think your remark very just : using the word in some measure in a different sense from which it is derived appears to me an imperfection in this mode of classification.

EMILY.

Families are to genera, what genera are to species; or, to follow up the comparison you made between plants and the human species, we might say, that families of plants were like nations of human beings; and that all these families collectively form the vegetable kingdom, in the same way as all the nations of the earth form the population of the world.

MRS. B.

Exactly so ; and the families of plants, like the different nations of the world, have each their peculiar characters and habits. Thus, independently of the analogy between their organs, plants of the same family often resemble each other in their mode of life, and in their peculiar properties. For instance, all the *Ficoideæ* have succulent leaves, suffer from moisture, and inhabit climates where the sun's rays are powerful; then, the *Malvaceous* family bear leaves of an emollient nature; the embryo of all the *Euphorbiaceæ* is of an acrid nature ; the roots of the *Valerianeæ* have all a particular smell, and act in a peculiar manner on the nervous system; the Cruciform family is, in all its branches, antiscorbutic. In a recent voyage, undertaken with a view of discovering the spot where the celebrated La Peyrouse was shipwrecked, the whole of the crew was afflicted with a scorbutic complaint, which was greatly relieved by feeding on an unknown plant of the cruciform family growing on the coast of New Holland — a remedy which was pointed out to them by the botanist attached to the expedition. There is another point of resemblance between plants of the same family, I have before mentioned; which is, that these alone are susceptible of being grafted on each other.

CAROLINE.

These analogies are extremely curious ; and I understand now, perfectly, how much superior your method is to those which merely indicate the name : for when I know that a plant has four stamina, it teaches me nothing further; whilst the knowledge that a plant belongs to such or such a family makes me acquainted, in a great measure, at least, with its structure and its properties.

H 2

EMILY.

I am glad, Caroline, that you are come round
to my opinion; for I felt a sort of instinctive con-
viction that it was in analogies of this description
that the interest of the study of Nature consisted.
But pray, Mrs. B., is it not possible to group
families together in the same manner as you have
done genera?

MRS. B.

Yes; there is still another step in classification,
which brings us to the three great distinctions
with which you are already acquainted.

First. That class of vegetables called *Dicotyledons*, relative to their organs of re-production;
and *Exogenous*, relative to their organs of nutrition.
This is by far the most numerous of the three
classes, comprehending about two-thirds of the
vegetable kingdom.

Secondly. The vegetables called *Monocotyledons*,
or *Endogenous*, according as you allude to their
re-productive or nutritive organs.

Thirdly. The class called *Acotyledons*, from
being destitute of cotyledons, are also called *Cellular*, because their nutritive organs have no vas-
cular system. To complete the comparison we
have followed up, these three great classes may
be considered as the three great continents of the
world; the different families of plants as the

various nations into which these continents are divided; the genera represent the families of each nation; and the species must be considered as the unity of the scale. This comprehends the whole system of classification, which every day becomes more extended and more perfect.

EMILY.

Let me see whether I can make out this genealogical table of plants.

The three grand divisions give birth to 250 families.
These 250 families produce 5000 genera.
And the 5000 genera, 60,000 species.

MRS. B.

The species, in their turn, give rise to races, varieties, and variations; but we shall not enter upon these subdivisions at present, as they are the result of an artificial, rather than of a natural, mode of propagation; and, indeed, their numbers are both too great and too variable to be reckoned in a table of classification. There are, for instance, no less than fifteen hundred varieties of the vine, and five hundred of the pear-tree: it is true that other plants do not afford so great a number.

CAROLINE.

If they did, you might almost as well undertake to count the individual plants as to number them.

But do not the number of species also increase by the discovery of new plants ?

No doubt they do. Since the death of Linnæus, about fifty thousand new species have been discovered, making, on an average, one thousand species every year.

The numbers, therefore, which I have given you, are intended only to enable you to form a general idea of the present state of the vegetable kingdom ; but they cannot be considered as permanent.

This explanation will, I hope, enable you to understand the basis of the *natural classification*, the details of which can be acquired only by study and practice.

But how can we study this system, since the English botanists follow that of Linnæus ?

Generally they do, but not exclusively. We already possess two excellent English works: the one called the *Flora Scotica*, by Mr. Hooker, in which the plants are classed both according to the system of Linnæus and that of Jussieu ; the other, the *British Flora*, very recently published by Dr. Lindley, Professor of Botany in the University of London, in which the plants are arranged according to the natural method. You may afterwards consult works of a more general description, which will carry you still further; and, when once you are accustomed to investigate the affinities of plants, your eye will enable you to guess, as it were, a great portion of what remains to be learnt.

EMILY.

If these affinities are so evident to the eyes of a botanist, whence comes it that they have only been so recently studied?

MRS. B.

Botanists have long been acquainted with the affinities of plants growing in great numbers in Europe. Thus, ever since botany has become a science, the Cruciferous, Gramineous, Umbelliferous families, and several others, have been distinguished. But those families which are dispersed over every quarter of the globe could not be classed until the analogy of the different plants had been discovered and studied; and travellers, ignorant of botany, are incapable of recognising the affinities of plants they have never studied.

EMILY.

What an interesting study the comparison of plants of different countries must be!

MRS. B.

Undoubtedly it is. This study is called bo-

H 4

tanical geography; and, if you wish to acquire some idea of it, we will make it the subject of our next conversation.

EMILY.

With the greatest pleasure.

CONVERSATION XXIV.

ON BOTANICAL GEOGRAPHY.

MRS. B.

At our last interview I promised to give you some idea of the laws which appear to regulate the distribution of plants on the surface of the globe.

CAROLINE.

Yes; this is the study you called Botanical Geography.

MRS. B.

It is a science of very recent date; indeed, it is only within the last few years, that it has been cultivated with any degree of success. It is founded entirely on the distinction made between the *habitation* and the *station* of plants.

CAROLINE.

I do not understand what you mean by this distinction. Are there two modes of indicating the country of a plant and the spot in which it grows?

MRS. B.

Precisely so. For instance, when you say that the tulip-tree grows in America, you point out what, in botany, is called its *habitation ;* when you say that it grows in marshy districts, you intimate its *station.* Thus, the term habitation relates to the geographical distribution of plants on the face of the globe, while station denotes the peculiar localities in which they are generally found.

EMILY.

I understand your meaning perfectly ; but I cannot conceive that any degree of importance can be attached to this distinction.

MRS. B.

I will explain it. You will readily admit that the nature of the soil, the aspect, the degree of moisture, &c., is sufficient to account for particular plants growing in certain spots rather than in others. Their *station* is thus explained by physical laws, with which we are more or less acquainted. The causes of their *habitation* are, on the contrary, perfectly unknown to us. Were you, for instance, to find in America (a circumstance not at all improbable) a marshy district, perfectly similar both in regard to temperature, moisture, and the nature of its soil, to another marshy district in Europe, the two marshes would be peopled with plants of a very different description. The cause of this singular phenomenon appears, therefore,

2

to have existed prior to the actual state of the globe, and is consequently impossible to explain.

EMILY.

It is true that the tulip-tree, of which you were just speaking, grows very well when transplanted to Europe; and I have heard that our walnut-trees thrive equally well in America: but neither of these trees grow spontaneously out of their natural country, or, as you call it, their habitation.

MRS. B.

Well, then; botanists, after having studied the surface of the earth under this point of view (as far as their imperfect knowledge of barbarous countries would admit), have divided the globe into twenty districts, which they named *botanical regions*. Each of these regions possesses a vegetation peculiar to itself, plants of the same species being seldom found growing (naturally I mean) in different regions.

CAROLINE.

How are these regions to be distinguished from each other?

MRS. B.

Those whose limits are the most correctly determined are separated from each other by a vast expanse of sea.

CAROLINE.

Why a vast expanse?   Would not a narrow

H 6

sea, like the Mediterranean, serve to define the
limits equally well?

No; narrow seas do not constitute a limit to
botanical regions. There is scarcely any difference between plants which grow in the north of
France and those growing in England, or between
the plants on the two opposite shores of the
Mediterranean. Nor do islands in the vicinity of
continents constitute a boundary, as they have
generally the same species of vegetation as the
neighbouring continent; while islands situated at
a considerable distance from continents have often
quite a different vegetation. For instance, the
plants which grow naturally in St. Helena and the
Sandwich Isles, are almost all different from those
of any of the continents.

Then I conclude that large tracts of continent,
also, must differ in the nature of their vegetation.

It is so in general; but as the old and the new
world approach very near to each other, if they
are not actually united towards the north pole, the
plants of the northern regions are nearly the same
in the three continents; and the further you recede
from the pole, the more distinct the different
regions become in regard to vegetation.

4

EMILY.

Are there any natural limits which separate different regions in the same continent?

MRS. B.

There are, but they are less defined than those separated by seas; so that there is a greater mixture of plants in these regions. Their natural limits in continents are, for instance, either extensive sandy deserts, such as those of Sahara, which separate northern Africa from Senegal, or chains of high mountains, which oppose an insuperable barrier to the conveyance of the seed by natural means; or, again, vast salt-plains, which prevent the germination of seeds.

EMILY.

But are there not a variety of means by which plants may be conveyed from one region to another?

MRS. B.

No doubt; and that accounts for plants appertaining to different regions often being found growing in the same. Rivers, for instance, and high winds, convey seed from one country to another; birds of passage transport the seed on which they feed; animals carry them in their woolly or hairy coats; and, finally, man conveys seeds wherever he goes: sometimes voluntarily, as corn and potatoes, which he has disseminated all over the known world; at other times, unin-

tentionally.   And it is owing to this casual transport, that the plants, and even weeds, of most of
our villages, have found their way to America.

CAROLINE.

Like Robinson Crusoe, when, by shaking the
dust out of a bag, he produced a crop of corn.

MRS. B.

Very true; but men have even gone further,
and conveyed seeds from one part of the world
to another, much against their intention or in-
clination; such as the seeds of the wild poppy
and corn-flower, which can never be completely
separated from the grains of corn.

But, independently of these emigrations, it must
be confessed, that there is a small number of similar plants existing in different regions, without the
possibility of explaining how they could have been
conveyed from one region to another.

CAROLINE.

This is quite a new idea to me: I always thought
that a great number of the same plants were to be
found in countries very distant from each other. I
have heard of the American elm, the apricot of
St. Domingo, and many other plants bearing the
same names, both in Europe and in America.

MRS. B.

This is owing, in a great degree, to the first

colonists who settled in America being ignorant of botany, and giving European names to plants, which, in fact, were very different from those whose names they assumed.

EMILY.

I suppose they considered it as a sort of tribute paid to their native country ; just as they gave the names of New York and New Holland, to countries very different from those of Europe.

MRS. B.

Another reason may also be alleged. It often happens that different species of the same genus inhabit different regions; for instance, the *Vaccinium Macrocarpum*, which we call Canadian cranberry, is of a different species from the *Vaccinium Oxycoccus*, or English cranberry, which we eat dressed exactly in the same way. Thus, also, the oak, the pine, and the maple, of the United States, are of a different species from those which bear the same name in Europe.

EMILY.

It appears, then, that there is no sort of connection between the classification of plants and their geographical distribution.

MRS. B.

There is some slight connection, but it is so variable that it is little to be depended on. Thus while certain families and certain genera are dis-

persed all over the world, others are confined to a
single region; all the *Cacti,* for instance, come
from America; the *Aurantiaceæ* from India or the
neighbouring countries; the *Epacrideæ* from New
Holland; and, amongst the genera, there are many,
every species of which inhabit the same region.
Thus, all the *Cinchonas* are derived from South
America; the *Gorterias* from the Cape of Good
Hope, &c.

It often happens that different genera bear so
near a resemblance to each other, that the various
species of the same genera or families are divided,
as it were, between them. For instance, a portion
of the *Pelargoniums* are situated at the Cape of
Good Hope, while another portion of the same
family grows in Van Dieman's Land. Botanists
have of late paid great attention to this subject;
but the results of their researches can be considered only as temporary, as it will ever be liable
to change so long as unknown plants remain to
be investigated.

EMILY.

Cannot you give us some idea of the result of
their researches?

MRS. B.

It has been calculated, for instance, that in almost all the botanical regions of the world one
sixth of the plants are monocotyledons; and that,
in regard to the other two classes; the number of
dicotyledons increases as you approach the equator, and that of the acotyledons, on the contrary,

as you draw nearer towards the pole. This rule does not prevail in islands situated at a great distance from any continent: in these the proportion of monocotyledons is greater, and that of the dicotyledons less, than is usually found in continental regions of the same latitude.

CAROLINE.

You do not mean to say that the same proportion of monocotyledons exist in Europe as in Asia — in cold northern countries as in tropical climates? To judge from the views I have seen of India, the greater part of the trees are of the family of Palms.

MRS. B.

A few of such magnificent trees make a great show in a landscape; but recollect that all our corn and grasses are monocotyledons. The difference between this class in England and in India consists, not in the number, but in the size, of the plant. The vigorous vegetation of tropical climates produces monocotyledons of stupendous dimension, while the chilling temperature of northern regions checks their growth; and if we go beyond the gramineous family, it is but to produce lilies, tulips, hyacinths, and other imperfectly-developed bulbous roots. It is only in the most southern parts of Europe that a few straggling palms denote the approach to a more vigorous region of vegetation.

CAROLINE.

But tropical climates produce corn and grasses as well as palm-trees.

MRS. B.

True, but in much less quantity; herbaceous plants require less heat and more moisture than is to be met with in such climates. The number of ligneous, compared to that of herbaceous plants, universally increases as you approach the equator.

EMILY.

Does this increase and decrease proceed in a regular progression from the equator to the poles?

MRS. B.

No: the number of annual plants, for instance, is very considerably greater in temperate than in either the tropic or frigid zones. The delicate structure of those plants render them incapable of resisting either the dry heat of the tropics or the severe cold of the polar regions.

EMILY.

We have also the advantage of the most beautiful and delicate colours in the vegetation of spring; while I have heard that, both in the polar and tropical regions, the spring-leaves are of a much darker and more sombre colour.

MRS. B.

Now that you have acquired some idea of what is meant by a botanical region, let us observe how the plants are distributed in one of these regions, and why different plants prefer different localities.

You will easily understand, that every plant, according to its particular structure, requires the concurrence of many circumstances in order to be brought to perfection.

EMILY.

No doubt. It is evident that the same soil, or the same degree of heat, light, or moisture, cannot be equally good for all plants.

MRS. B.

When plants shed their seed, it is more or less dispersed by wind, rain, or other natural agents, and is finally deposited on a soil which may or may not be favourable to its germination. Thus, in particular spots, a sort of struggle takes place among the different species of vegetables which it produces. The most vigorous plants, and those best suited to the nature of the soil, make the greatest progress, and ultimately exclude the others.

CAROLINE.

So that in the vegetable, as well as in the animal kingdom, the strong oppress the weak, and a contest takes place even among flowers, to all appearance the symbols of peace and harmony.

MRS. B.

I am sorry to spoil your poetical ideas of vegetation; but such is the law of Nature. You will
now understand, that the richer the soil the greater
is the number and variety of plants that can grow
in it. Thus, in tropical climates, the forests are
composed of a much greater variety of trees than
in the temperate zone; and, as you approach towards the polar regions, the number of different
plants gradually diminishes.

EMILY.

It is, perhaps, on this account that in the high-
lands of Scotland we meet with immense tracts
where no plant is to be seen growing but heath or
furze.

MRS. B.

Precisely. These species of plants being of a
hardy nature, and able to live in a soil from which
most other plants are excluded, meet with no competition, and establish a colony apart from other
plants. Such plants are called by botanists *Social*,
from their habits of living together in societies.

CAROLINE.

I think they should rather have been called un-
social, from their excluding plants of a different
species.

MRS. B.

They at least deserve the name of inhospitable :
the *Potamogetons*, which grow in stagnant waters,

Kelpwort (*Salsola*), and Saltwort (*Salicornia*), which grow in salt districts, are of this description. There are some plants which become social from their mode of propagation; those, for instance, which have spreading roots, such as the *Hieracium Pilosella*, or Mouse Ear Chickweed. Plants, on the contrary, whose seeds are crowned with a tuft, which enables the wind to have more power over them, are dispersed to a great distance : between these two extremes there exists a great variety of intermediate degrees.

There are some plants which, so far from excluding those of a different species from their society, seem to take delight in the neighbourhood of trees to which they themselves bear no resemblance : thus, the *Salicaria* loves to grow at the foot of the willow; the *Monotropa*, at the foot of the pine ; the *Saxote*, to grow amongst oats.

EMILY.

What can be the reason of this singular kind of attachment of one species of plants for another ?

MRS. B.

Several have been assigned : first, that plants of different species frequently require the same soil; the next (of a more doubtful nature) is, that the exudations of some plants may promote the growth of others of a different species ; a third reason alleged is, that certain plants often serve to protect others of a different species, as hedges and bushes protect the creeping plants which grow between their branches.

EMILY.

It appears, then, that we can in some degree explain that prodigious mixture in the vegetable kingdom, in which at first I thought there was no sort of order.

MRS. B.

There is always order in the works of Nature; and what appears to us disorder is the result of different laws acting at the same time. By following the mode of reasoning I have pointed out to you, and by constantly comparing the structure and the habits of plants with the nature of the soil in which they grow, a great number of curious facts may be explained. I am glad to have drawn your attention to this subject; it will be a source of amusement in your walks: and the greater the number of plants you become acquainted with, so as to be enabled clearly to distinguish their different species, the more interesting will your observations prove.

CONVERSATION XXV.

MRS. B.

LET us now examine to what extent the natural state of plants can be modified by the art of man. For this purpose it will be necessary for me to make you acquainted with certain differences which exist in plants of the same species.

A species, you recollect, comprehends all those plants which bear so great a resemblance to each other that we may reasonably suppose them to be descended from the same parent stock. But, independently of this general similitude, each species admits of various shades of difference, some of which are strongly marked, and of a permanent nature; others more slight and evanescent: hence spring the three modifications of *Races, Varieties,* and *Variations.* Several *races* derive their origin from the same species; and the points in which they differ are of so decided a character, that they are continued from the parent-plant to its offspring, or, in other words, when it is propagated by seed.

Varieties are a subdivision of races; in which the points of difference are of so slight a character, that they are continued from one individual to another only when the plant is propagated by subdivision; that is to say, by grafting, budding, or layers, but are obliterated when it is raised by seed.

Variations are the feeblest of all deviations: they originate in the peculiar ·circumstances or situation of the plant, such as peculiarity of soil, temperature, &c., and are susceptible of being continued to successive individuals only if placed under similar circumstances.

Now, the art of man has great influence in varying and multiplying these several modifications of species. If, for instance, the pollen of the flower of one species be made to fall on the pistils of another species, one of two things may happen: either the flower will produce no seed; or, if it produce seed, the plant which results from it (which is called a Hybrid) will partake of the form and nature of the two plants from which it springs; and hybrids very rarely produce any seed.

CAROLINE.

It is then, I suppose, only performed as a curious experiment, since the seed is lost, and nothing is gained in exchange.

MRS. B.

True; but the result is very different, if, in two plants of the same race but of different varieties,

the pollen of one be made to fall on the pistils of
the other, the blossom will in general bear fruit,
and thus a new variety will be produced, differing
from those from which it drew its origin. Let us
suppose, for instance, that there were but two
varieties of cabbages in nature, the one spherical,
the other spreading : by the intermixture of the
pollen of these two, a third variety would be produced; and by continuing the process between
these three varieties, ten, twenty, or thirty new ones
would result. But as these varieties bear seed capable of reproduction, it is, in fact, new races which
are formed.

In Belgium, the horticulturists, with the most
patient perseverance, produce, by this process, a
great number of new varieties of fruit trees, which
they propagate by seed, and thus give birth to
new races; but this is extremely tedious, for it is
many years before the fruit tree raised from seed,
is capable of bearing fruit.

EMILY.

This period might be accelerated by grafting;
but then that process would alter the nature of
the new variety of fruit.

MRS. B.

Certainly; the Dutch are celebrated for the
beauty, or rather the variety of colour, of the
tulips they have thus introduced. These flowers
change their colour during the first seven years,

they afterwards never vary: this renders a course of experiments, with a view to produce certain colours permanently, much more tedious, and, consequently, more expensive than with most other plants; and the Dutch horticulturists prosecuted their labours with such enterprising zeal, and the passion for flowers was, in that country, carried to such excess, that it was thought requisite to enact a law, forbidding the sale of a tulip for above the sum of four hundred pounds.

EMILY.

Is it possible that any one would go to so great an expense for a simple flower! It is by these means, I suppose, that many fruits and flowers have of late years been so much improved. The great variety of beautiful geraniums and gigantic strawberries are, doubtless, the result of similar experiments; but the flavour of the fruit does not, I think, correspond with its size; I even doubt whether the bulk is not increased at the expense of the flavour.

MRS. B.

It will seem to be diminished to the palate, if the same quantity of flavour be diffused over a greater bulk of fruit; but I believe that the horticulturists consider that they have improved the flavour, as well as the size of the fruit.

The influence of culture on variations results from its influence on the soil, and the quantity and quality of the nourishment afforded to plants.

Hence some parts of a plant may be made to prosper more than another; the stem more than the foliage and fruit, if timber be required; the leaves more than the seed, if grasses; or the fruit more than the leaves, with most fruit trees. A change of colour may also be produced. Thus the Hydrangia, when first brought from the Isle of Bourbon, was blue; in this country it is commonly of a pale pink, and it is the soil principally which has effected this change; for if cultivated in a ferruginous soil, similar to that of its native land, the blue colour is reproduced. Pink flowers may be thus changed to blue or white; but cannot be made to assume a yellow colour; thus the Hydrangia or the Campanul may be varied from pink to blue or white, but you never see them of a yellow colour.

EMILY.

That is true; Hyacinths are also pink, blue, or white, but they are seldom of a yellow colour.

MRS. B.

They are the only flowers which form an exception to the rule, being sometimes yellow.

The neighbourhood of the sea produces a variation in plants, rendering them more succulent or fleshy.

Grafting also modifies the variations of plants. The art of pruning has very considerable influence, by modifying the direction of the sap;

but its effect, however great on the individual
plant, produces no change on its successors.

Trees are pruned with a view to improve their
beauty, their health, or their produce.   Trees were
formerly cut and trimmed into all kinds of grotesque figures, according to the tasteful ideas of
beauty of our ancestors.   Since this barbarous system has been exploded, that of heading young
trees, in order to thicken the branches and foliage,
has been introduced; but this, we have agreed,
injures the natural port and character of a tree;
and all that is allowed in the present times, in
order to improve the appearance of a tree, is to
strip off the lower branches, in order to prevent
its assuming the form of a bush.   This operation
should not be performed too soon: the stem, while
young, requires the aid of these lower branches
to carry on the process of vegetation, and supply
it with nourishment: they pour their cambium
into the stem at its base, and thus assist in increasing its vigour.

EMILY.

Yet, would not this operation become dangerous, if long delayed? for the larger the lower
branches are suffered to grow, the more serious
will be the effect of their amputation.

MRS. B.

The proper time for lopping them is, when the
tree has attained sufficient vigour to enable it to

recover from the wounds, in the course of the year.

Resinous trees suffer from pruning, by losing too much of their resinous juices : fir trees should never be pruned ; but if planted in groups, as we see them growing naturally, the lower branches, being deprived of light and air, dry up and perish : it is thus that Nature prunes them without the infliction of a wound, from which the resinous juices would flow, to the great detriment of the plant.

In regard to the pruning, which relates to the health of plants, not only should all the dead branches be carefully removed, but the pruning knife must penetrate into the quick of the wood. It is advisable, also, to cut away all the parts which are diseased, as these seldom recover, and would continue, during a few years, of sickly existence, to absorb, uselessly, a portion of the sap, and very probably, during this period, to communicate their malady to the contiguous branches.

All branches seriously injured by hail, should be immediately removed; they will then rapidly shoot afresh, and, in the course of a few weeks, their loss will not be perceived.

EMILY.

Greenhouse plants must require a great deal of pruning, for as their roots cannot grow freely in search of food, the branches must be diminished,

in order to correspond with their limited quantity
of nourishment.

True; both root and branch require pruning
annually, when the plants are fresh potted. But
observe that the gardener takes care to atone, as
far as lies in his power, for the contracted sphere
in which they vegetate, by affording them as much
food as can be contained in so limited an extent of
soil.

Pruning fruit trees is done with the view of
either increasing the quantity, or ameliorating the
quality, of the produce. It consists in retarding
the descent of the cambium, in order that by remaining longer in the branches, it may nourish
them more abundantly. For this purpose, the
branches which grow vertically should be pruned,
because the sap, descending through them straight
downwards, moves with greater velocity than
when it descends obliquely, as it does in lateral
branches.

It has sometimes been found advantageous to
bend down the vertical branches, in order that the
cambium should be compelled to rise, in its return
from the extremity of the branch; and the time
required to overcome this difficulty retards its
march, and enables the branches to absorb more
nourishment from it, during its passage.

Espaliers are usually trained in the form of a
fan, by cutting away the central stem: or the
stem may be preserved, provided that the branches

be trained laterally; for it is in these, rather than in the stem, that it is essential to diminish the velocity of the cambium.

You recollect my having already made you acquainted with three species of buds: those which produce fruit; those which develope leaves only; and those of a mixed nature, containing both fruit and leaves.

CAROLINE.

Yes; and we observed that the more fruit buds escape the pruning knife, the greater will be the crop of fruit.

MRS. B.

Care should be taken, however, not to leave more fruit buds on the tree, than the sap will be able to bring to perfection, else the quality of the fruit will be deteriorated. Good gardening consists in preserving as many fruit buds, as the tree can nourish without exhaustion; for if you force a plant to labour beyond its strength, either the fruit will not ripen, or its size and flavour will suffer.

CAROLINE.

But this pruning, with a view to improve the quality of the fruit at the expense of the quantity, is an unnatural state of vegetation, which, I should suppose, would eventually be prejudicial to a tree.

MRS. B.

I cannot consider it so: the finest trees and the choicest fruit, are those in which art has judiciously assisted and modified the efforts of Nature. We contribute to the health and general prosperity of the tree by preventing it from bearing an excess of fruit; and we make amends for the diminution of quantity by the increase of its size and flavour.

CONVERSATION XXVI.

ON THE DEGENERATION AND THE DISEASES OF PLANTS.

MRS. B.

WE shall preface the history of the diseases of plants by that of the degeneration of their organs, which often undergo a species of metamorphosis, and, instead of being developed in the usual manner, degenerate into monstrosities.

There are several causes which produce this effect on plants: 1st. The natural soldering, or cohesion, of the parts. You frequently see the leaves of branches, the petals of flowers, and even fruits which unite, forming double leaves, double flowers, and double fruits.

Such cohesion sometimes regularly occurs. The single petal which forms the corolla of many flowers, such as the convolvulus, is composed of the union of several others; but as it is not unfolded until after the junction is completed, we are led to consider it as a single petal; and such flowers are called in botany monopetalous.

I 5

EMILY.

But where this union regularly occurs, it should, I think, be considered as the natural state of the plant, and not as a monstrosity. Pray, how does it take place? Is it a species of grafting one petal upon another?

MRS. B.

No; it is rather a simple adhesion than a continuity of vessels through which the sap passes. The petals in which this adhesion so frequently occurs have no liber; and this, you know, is essential to the process of grafting, as it is through the vessels of the liber that the cambium descends.

Another species of monstrosity arises from a want of vigour in the plant to bring all its parts to maturity. That which most commonly fails is the seed, which is produced in such abundance, and requires so much nourishment to ripen, that the greater part perishes in the bosom of the flower. The blossom of the horse-chesnut, for instance, contains six seeds, enclosed in three cells; but one only, or at most two, come to maturity. It is the same with the oak; it has six seeds, but only one acorn is brought to perfection.

CAROLINE.

And to what cause is the want of developement owing? If the plant be incapable of ripening so many seeds, why has Nature furnished it with so useless an abundance?

MRS. B.

The causes of these abortions are probably numerous; but the principal one is, no doubt, a deficiency of nourishment. Yet so far from inferring that such failures imply a want of regularity in the laws of Nature, it is to them that we are indebted for one of the most efficient means of ascertaining the order which reigns in the natural world.

A third species of monstrosity results from a degeneration of the organs, which disables them from fulfilling the purpose for which Nature originally designed them. Thus, in some plants, the leaves do not sprout, and the stem, receiving the nourishment which the leaves should have absorbed, swells out to a considerable size, and expands like leaves. The Xylophylla and the Cactus opuntia are constantly in this state. It is said, that the leaves of these plants bear flowers; but the fact is, they have no leaves; the flowers grow on the expanded stems.

Flowers having double blossoms are also classed among the tribe of monsters. This arises from the stamens being too abundantly nourished. They swell out, flatten, and are converted into petals; hence the flower becomes double. Thus we have double roses, double stocks, double blossom cherry, &c. The process of this metamorphosis is very plainly discernible in the double hyacinth and the double tulip, where many of the stamens are completely transformed into petals:

others, while expanding for that purpose, still partially retain their original form.   As this metamorphosis never occurs but when the anthers have perished, it is probable that they are starved by the stamens absorbing the whole of the nourishment.

EMILY.

It is, I suppose, owing to the destruction of the anthers, that double flowers bear no seed. But why should such beautiful productions of Nature be stigmatised by the name of monster? It is considering beauty as a deformity.

MRS. B.

However disagreeable are the ideas commonly annexed to the term monster, the word simply implies a deviation from the common course of Nature. In the animal kingdom, such a deviation almost always excites disgust, and is associated with the idea of ugliness. Were there consciousness in plants, they might very possibly consider the unusual quantity of petals and the deficiency of anthers as a deformity; but we, who look upon a flower merely to delight our sight with its form and colour, associate the idea of beauty to this unnatural state.

Another instance of degeneration is, when the petioles or foot-stalks are transformed into leaves. The Acacia, for instance, has six or eight pair of leaves, a number which diminishes every year, till at length the foot-stalk is wholly deprived of leaves ; but receiving all the nourishment which was previously distributed to them, it expands, flattens, and is itself finally converted into a blade, resembling a leaf.

EMILY.

Though the acacia is not a very common tree in England, I have seen a great number on the Continent, but never observed the species of metamorphosis you describe.

MRS. B.

The acacia to which I allude is that of Arabia, which produces gum arabic, and is known in Europe only as a hothouse plant. It is the original and only true acacia. The tree we cultivate under that name is derived from North America : it obtained the name of acacia from some resemblance between its fruit and that of the Arabian plant, and was distinguished from it by the title of *false* acacia : but as the American tree multiplied in Europe whilst that of Arabia was known only to horticulturists, the epithet false was dropped, and it now usurps the name which really appertains to the Arabian plant. Its botanical name is *Robinia.*

Instead of wholly disappearing, folioles often degenerate into tendrils, for want of sufficient nourishment. The flower-stalk, or peduncle, is also sometimes converted into tendrils. This occurs constantly in the vine. The plant at first shoots out abundance of large leaves and clusters of

grapes, when, after a time, the food proves insufficient to support such a profuse vegetation; the new
leaves, gradually unfolded, are of smaller dimensions, and the clusters of grapes contracted in size.
Still nourishment is wanting, and the later shoots,
incapable of developing either flower or leaf, are
converted into tendrils. Is this an imperfection in
the system of vegetation, or is it not rather a
beautiful contrivance, to enable the plant, when it
has sprouted all the branches it can nourish, to
sustain these branches by means of the tendrils in
which they terminate, and which cling to the first
object capable of affording them support?

EMILY.

These organs, which you call degenerated, appear to me to serve a purpose no less useful than
the functions they would have performed had they
come to a state of perfection. But do all the
various sorts of tendrils of climbing plants result
from the degeneration of other organs?

MRS. B.

There is great reason to suppose so. The most
common of these degenerations is the transformation of the young shoots of branches into thorns.
When a plant shoots more branches than it can
nourish, the most weakly almost wholly cease to
grow. The scanty sustenance they receive serves,
however, to harden and strengthen them : hence
the tender extremity is converted into an indurated

sharp point, capable of inflicting wounds, which
you must often have experienced.

CAROLINE.

Is it not singular that these two last degenerations, resulting from a similar cause, should be so
different in their effects?   In the thorn the food
hardens without extending the shoot, whilst in the
tendril it is extended to a considerable length, and
is extremely flexible and slender.

MRS. B.

Nature has so contrived it (though by means
which are unknown to us), no doubt, with a view
to provide support for climbing plants, which are
too weak to bear the weight of their produce ; and
where no such assistance is required, she has converted the abortive shoot into an arm of defence.

EMILY.

Would, then, these plants have fewer tendrils
and thorns, if transplanted into a richer soil?

MRS. B.

No doubt ; because a greater number of young
shoots would be brought to perfection.   M. De
Candolle transplanted a wild medlar-tree, covered
with thorns, into his botanical garden, and in the
course of three years not a single thorn was to be
seen upon it.

EMILY.

Yet I have never observed that the rose or the

gooseberry bush lost any of their thorns by cultivation.

MRS. B.

They are not thorns, but prickles, which grow
upon the rose, the bramble, the gooseberry, and
many other plants; and these are quite of a different nature. The prickle is a natural appendage,
which has no connection with the wood; it springs
from the bark, and is peeled off with it; and since
it does not result from the degeneration of any
organ, it is not susceptible of being diminished by
cultivation.

The peduncle of the grape terminates in a
tendril, when the vine is loaded with as many
clusters of fruit as it can bring to maturity. But
in a very favourable soil, more grapes would be
produced, and this transformation of the fruit-stalk
takes place later, and probably less frequently.

CAROLINE.

And may not monstrosity of organs be produced
by plants having too much nourishment?

MRS. B.

Certainly; it happens, if the nourishment, instead of being equally disseminated throughout the
plant, partially increases the growth of any particular part, all disproportion of size among the relative parts, is a deviation from the regularity of
Nature, and must be considered as deformity; but
as it is much more common for plants to be under

than over fed, the monstrosities which arise from the latter cause are of rare occurrence.

Though these various irregularities and metamorphoses are classed under the head of monstrosities, I am far from considering them as evils: I view these changes as advantageous to plants, and if naturalists rank them as imperfections in the system of vegetation, they are, by the beneficence of Providence, turned to such good account, that we cannot but estimate them as blessings.

We shall now proceed to consider the influence of culture on the diseases of plants. The botanical physician must not rest satisfied with studying the symptoms of a disease, for the same symptoms may be produced by very opposite causes: thus, plants turn yellow if they receive either too much or too little water; and, in order to afford a remedy, the cause of the malady must first be carefully investigated.

CAROLINE.

And that must be very difficult: since, in examining the patient, you cannot ask him any questions.

MRS. B.

Fortunately, the diseases of the vegetable kingdom are of a less complicated nature than those of animals.

The diseases of plants may be ranged under six different heads: —

1. Constitutional diseases.

2. Diseases arising from light, heat, water, air, and soil, improperly applied.

3. Diseases arising from contusions and external injury.

4. Diseases occasioned by the action of animals on plants.

5. Diseases proceeding from the action of vegetables on each other.

6. Diseases arising from age.

Variegated or party-coloured leaves, such as those of the box and the holly, are classed as constitutional diseases. They arise from certain juices of plants, which, from some unknown cause, change their nature, and thus affect the colour of the leaf. These changes are preserved if the plants are multiplied by subdivision, and even sometimes continued when propagated by seed.

The second class of diseases results from circumstances connected with the undue supply of elements, which are in themselves necessary to vegetation; such as temperature, light, water, air, soil, &c. If a plant has too much or too little light, heat, or water, it has no means of avoiding the excess, or of compensating for the deficiency.

CAROLINE.

The poor plant, it is true, rooted to the ground, cannot, like an animal, fly the evil, or seek a remedy; it must patiently submit to it, and endure the diseases it entails: if the soil afford too much nourishment, it must continue feeding, and cannot stop when its appetite is palled.

EMILY.

Or, what is worse, and more frequently the case, when the soil does not yield a sufficiency of nourishment, it cannot seek it elsewhere, and famine must debilitate the roots, and diminish that vigour which would enable them to stretch out their fibres over a greater extent of soil.

MRS. B.

We have already entered so much into detail on the influence of light, heat, water, and soil, on plants, that I shall confine myself to recalling a few of the most essential points to your memory. Excess of light produces too much excitement; the oxygen escapes, and the carbon is deposited too rapidly; the plant vegetates in a fever, and the sap, incapable of supplying its wants, is exhausted; the plant withers, and the leaves fall off.   There are two modes of remedying this disease; either to increase the aliment, or diminish the vegetation; the first may be done by plentiful watering, the other by diminishing the intensity of the light.

Excess of heat dries up the juices; if you attempt to remedy this by plentiful watering, the plant sprouts leaves, but very little fruit.

CAROLINE.

This sort of vegetation must be well adapted to

meadows, where a produce of leaves is principally aimed at.

MRS. B.

True; but plentiful irrigation is not always attainable: where it can be had, no evil effects need be apprehended from the sun.

A deficiency of heat produces dropsy, and often rotting : the most delicate parts of the plant first begin to decay, such as the articulations of the branches and of the leaf-stalks; hence the leaves and young branches fall off. A plant evaporates much more water than it retains; it may be compared to a tube into which you introduce water : now, it is evident, that the more you pour in at one end, the more must be poured out at the other : the evaporation by the leaves must correspond with the absorption by the roots, else the plant will suffer.

Plants are also often injured by exposure to external moisture. Rain is more hurtful to the wood than to the bark; the latter is a sort of great coat, provided by Nature to shelter the wood from the inclemencies of the weather: she has stored it with carbon, to enable it to resist putrefaction; and with siliceous earth, to render it firm and durable: but if, as it sometimes happens, the great coat be rent and ragged, the rain penetrates into the wood (which is very differently organised), and having no means of escaping, the stem becomes rotten.

EMILY.

Among the injuries plants sustain from rain, we must not forget that of its making the pollen of flowers burst before it is mature, and hence preventing the seed from being brought to perfection.

MRS. B.

True; but we have already entered sufficiently into detail on that subject.

In regard to the influence of the air, I have formerly observed, that the agitation which the wind gives to plants is advantageous if not carried to excess: the cambium, being a thick viscous juice, requires motion to promote its descent.

CAROLINE.

Yet you have said that the great aim of the gardener is to retard the descent of the cambium, in order that, by remaining longer stationary in the branches, it may afford more nourishment to the fruit.

MRS. B.

That is true, if the production of fruit be the object aimed at; but if, on the contrary, it be timber, we must promote the descent of the cambium into the trunk, instead of endeavouring to detain it in the branches.

Gentle exercise is, however, generally advantageous in the vegetable economy, and promotes the circulation of the juices; while violent motion occasions either exhaustion or fever. Hence the objection to props and espaliers; which we have already noticed. Boisterous winds are also mechanically injurious to trees, rending their branches, and sometimes tearing up their roots from the soil.

Plants are affected by the nature of the atmosphere in which they grow. There is nothing more prejudicial to them than smoke.

EMILY.

I am surprised at that; for smoke, you have told us, consists of small particles of carbon which have escaped combustion; and carbon, you know, is the favourite food of plants.

MRS. B.

The particles of smoke, though apparently so small to our senses as scarcely to be distinguished when separate, are mountains compared to the very minute subdivision which matter must undergo in order to enter into the vegetable system. Smoke may clog the pores of plants, but can never gain admittance through them.

EMILY.

But smoke is always accompanied by a current of hot air, which must be strongly impregnated with carbonic acid; and in this state the carbon is so minutely subdivided as to be quite invisible, and, I suppose, sufficiently so to enter the pores of plants.

MRS. B.

If plants absorb carbonic acid by their leaves, or any part exposed to the air, it can be but in very small quantities. Under common circumstances, it enters into their system only by their roots; it is their leaves which decompose it. Carbonic acid gas is as prejudicial to plants externally as it is to animals; for plants, under a receiver containing carbonic acid, die in the course of a few hours. Azote and hydrogen do not appear to be injurious to plants, unless in such quantity as to diminish the proportion of oxygen in the atmosphere, which their vegetation requires.

Third class of diseases arises from contusions or other external injury.

The accidental loss of their leaves, from whatever cause it may proceed, must be considered as a disease of plants: if it is not the effect, it is the cause of one; for when the sap rises to the branches, and finds no organs to elaborate its juices, it descends almost in the same state in which it rose, a thin crude fluid, little adapted to the nourishment of the stem and branches. Under these circumstances, its only resource is to feed and develope young shoots, which Nature intended should sprout only the following year. The sap is then elaborated in the leaves of the new shoot, is converted into cambium, and the regular circulation is restored.

How wonderfully prolific Nature is in resources to remedy any accidental interruption to her regular progress ! One would almost imagine the sap to be endowed with a sort of instinct, when we find that it is no sooner disappointed in meeting with those organs requisite to its perfection, than abandoning its natural course, it busies itself in feeding and prematurely forcing into vegetation the organs which are deficient.

MRS. B.

This admirable fund of resources springs from an origin far superior to instinct. Its immediate cause is, it is true, probably either mechanical or chemical. The sap, for instance, cannot deposit the various juices required by the different organs, when a deficiency of leaves prevents these juices from being secreted. In its immature state it is, in all probability, better able to supply the elements required for the vegetation of buds; and thus the young shoots are prematurely forced into life. The mere mechanical philosopher will rest satisfied with this explanation; but if to the reflecting mind be added a feeling heart, he will discover that the beneficent Author of nature has so admirably regulated the laws by which it is governed, that they frequently find in themselves means of supplying remedies and resources against accidental contingencies.

CAROLINE.

This is, indeed, admirable. In a work of human

mechanism, however ingeniously contrived or skilfully executed, constant attention must be paid to
watch and remedy any accidental defect; whilst
the laws of Nature are of so perfect a description
that they are stored with those remedies which the
mechanist is obliged to supply.

MRS. B.

The loss of bark is so serious an injury as often
to prove fatal to plants. If the evil prevail entirely around the stem, so as to effect a complete
solution of continuity, the cambium can no longer
descend, and the plant must inevitably perish.

CAROLINE.

But do you forget, Mrs. B., that in cutting a
ring in the bark, to improve the fruit, you perform
the very operation you say is so dangerous?

MRS. B.

This ring, you must recollect, is so narrow, that
the swelling of the upper edge, from the accumulation of sap, soon produces a re-union of the
severed parts: but I was alluding to the destruction of the bark to so great an extent as to pre-
clude all chance of such a remedy. If the bark
be only rent on one side of the stem or branch, it
may be considered as a partial infirmity, of which
the plant may recover. For this purpose, the
diseased part should be carefully cut away, and the
wound be covered with an ointment. Let us sup-

pose the rent to be of a long oval form, as is
generally the case; the cambium, when it reaches
this spot, meeting with obstruction, will accumulate, and produce a swelling on the upper edge of
the wounded part : this will gradually descend on
each edge of the severed bark, till it meets at the
bottom, and the swelling will increase, till the two
sides unite, when the wound will be healed.

EMILY.

I have often observed the swelling of the bark
where a branch has been lopped ; but it remains a
protuberant ring around the wound, and does not
close, so that the central part of the wood remains exposed.

MRS. B.

In this instance, not only no ointment has been
used to shelter the part affected, but the wound
being of a circular form, it is more difficult for the
edges of the bark to meet.   The young wood,
however, which it is the most essential to shelter,
is covered by the swollen ring of bark.   The fla-
gellations which trees sometimes undergo to bring
down the fruit are injurious to them, by wounding
the young branches; it is so also to the fruit, unless
these be of the nut kind : for apples, pears, and
olives, when thus brought down, are bruised and
very liable to rot.

Plants often suffer from improper pruning.
When a tree is lopped of its branches, they should
be cut off obliquely; the sap, when it rises to the

wounded part, will then flow down its slanting surface, while, if the amputation be made horizontally, not only will the sap be less able to run off, but the wound will be more exposed to the rain and wind, and putrefaction will probably ensue.

EMILY.

I have seen the trunks of old willows, the branches of which are lopped every year, become perfectly hollow; which arises, no doubt, from the wood being thus injured.

MRS. B.

This operation, which is called pollarding a tree, is done with a view of turning the branches to the greatest advantage: in willows, generally, for basket-work; in other trees, for fuel.   When a tree is in the full vigour of life, it will be able to resist such merciless amputation; but when it becomes aged the wood will not support it without decaying.

Slight contusions, instead of being prejudicial to plants, produce an excitement which accelerates vegetation.   The prick or perforation of insects, which we have noticed in the fig-tree, simply occasions a small swelling somewhat analagous to that produced by a blow given to an animal: in this swelling a minute quantity of sap is deposited, which nourishes more abundantly, and, consequently, developes more rapidly, the surrounding parts.

196   DEGENERATION, ETC. OF PLANTS.

This leads us to the class of diseases arising from the action of animals on plants. But it is too late to enter upon it to-day : we shall reserve it for our next interview.

CONVERSATION XXVII.

THE DISEASES OF PLANTS CONTINUED.

————

MRS. B.

PLANTS suffer much from their leaves being de-
voured, either by quadrupeds or insects. The
former not only wound the branches in obtaining
the leaves, but, if the soil be of a loose nature,
they disturb the young roots; hence pasturage is
esteemed injurious in loose and wet soils. But the
insect tribe is a far more insidious and fatal enemy.
Insects not only perforate the plant, in order to
deposit their eggs, but, when these eggs are hatched,
the larvæ or grubs prey upon the plants which
have afforded them shelter, devouring their leaves,
and often rotting the wood by their acrid juices.
Most of these insects bear the name of *Cynips:*
that which produces the swollen excrescence called
gall-nuts, from which ink is made, is one of the
most remarkable. The smoke of tobacco and
washes made of infusions of that plant are the best
preservatives against these minute but inveterate

K 3

enemies.   The insect called Cochineal fastens itself
to the bark of trees, and sucks the juice through it.
The black spots on orange-trees are insects of this
class : they are generally pernicious in greenhouses,
and should be brushed off.

CAROLINE.

I thought that cochineal was of a bright red
colour, and that the insect was peculiar to hot
climates.

MRS. B.

The species you refer to comes from Mexico,
and feeds on the *Cactus Opuntia,* from which it
derives the name of cochineal ; but there are many
other species of this insect, which are not confined
to tropical climates.

The fifth class of diseases results from the action
of plants on each other.   Plants being destined by
Nature to produce a much greater quantity of seed
than they can possibly bring to maturity, we may
consider them as constantly struggling with their
neighbours to obtain nourishment for their numerous offspring: thus they impoverish each other,
and check that vigour of vegetation, which would
take place, had every plant sufficient space and
food not to interfere with the wants of its neighbours.

EMILY.

That is very evident in the superior vegetation
of a single tree, which has ample space for its
branches, and food for its roots, to that of a tree

in a crowded forest, where every inch of ground is disputed by surrounding plants.

MRS. B.

But, independently of this general competition for food, there are various other modes by which some classes of plants are noxious to others. Among these the parasitical plants stand pre-eminent. There are two classes of this description, distinguished by the epithets of *false* and *true*. The false parasite fixes itself to the plant, without feeding on its juices; while the true parasite feeds on the plant to which it adheres. These two classes are each subdivided into external and internal parasites, denoting the parts of the plant which they attack.

The false parasites consist of mosses, lichens, and fungi, which grow on living plants just as they would grow on a rock or a dead tree.

EMILY.

Such as the various mosses which grow on the stems of fruit-trees. But, if they do not feed on the tree, whence do they derive their nourishment?

MRS. B.

From the moisture of the atmosphere, and, possibly, from the relics of some preceding mosses, which supply a few particles of vegetable mould.

CAROLINE.

If they do not feed on the juices of the tree, in what manner do they injure it?

MRS. B.

Chiefly by attracting moisture to the stem, and thereby endangering the wood; and also by affording a lodgment for insects. In these temperate climates, however, the harm they do is not of a very serious nature; but, in tropical regions, parasitical plants grow with such luxuriance (the vanilla, for instance,) that the tree suffers mechanically from the weight of the mass it has to bear.

EMILY.

I recollect, in the Caschines of Florence, seeing many of the elm-trees so completely covered with ivy, that I at first sight concluded the tree itself was of that description.

MRS. B.

Ivy is a creeping plant, not a parasite. Its roots are planted in, and feed on, the soil: all it requires of the tree which it embraces, is support. Yet these plants, as you observe, are frequently prejudicial. I have seen trees whose branches have been so enveloped, and strangled, as it were, with creepers, that scarcely any room was left for its own proper foliage; and the growth of the tree was considerably impeded.

But to return to our parasites. The Rhizomorpha is a false internal parasite, which attacks wood; and, though it does not feed upon its juices, the mere growth of the plant proves fatal to it, disorganising its parts, and reducing the wood to a sort of vegetable mould. This malady seldom occurs but in very aged trees.

EMILY.

We, artificial beings, whose aim is to have plenty of sound timber for building, consider this as a dreadful malady; but, in the course of nature, it may, perhaps, simply be a means employed to reduce old or dead trees, to the state in which they are fitted to return again into the vegetable system, for this mould must afford rich food for other vegetables.

MRS. B.

In natural forests, where the hand of man does not interfere to turn the timber to his own account, it is certainly desirable that Nature should devise some means of hastening the decomposition of wood, a substance so hard and compact that it would require a great length of time to effect it by the usual process of decay. In this operation, the Rhizomorpha is aided by a tribe of insects, which take up their abode in the cracks and crevices, it has made in the wood.

CAROLINE.

The Mistletoe is, I suppose, a true parasite; for

K 5

it derives its nourishment from the tree to which it is attached.

MRS. B.

The seed of the mistletoe fastens itself to the tree by means of a glutinous substance with which it is covered.   The radicle of this seed shoots out in a manner different from that of any other plant: being too feeble, on its first entrance into life, to penetrate so hard a soil as wood, it shoots out in some other direction.

CAROLINE.

It grows, then, in the light and air, which must be an equally uncongenial soil !

MRS. B.

True; and it no sooner makes this discovery, han it changes its course, and, curving round, retraces its steps towards the branch whence it sprouted.

CAROLINE.

Just as if it were conscious that the soil it had abandoned, was that in which it was destined to grow.

MRS. B.

It is said that it is in order to avoid the light, that it alters its course; for roots, you know, dread the light as much as leaves and branches delight in it.

EMILY.

The dread of the former may, no doubt, be as mechanically explained as the delight of the latter.

MRS. B.

Certainly. The extremity of the root having now acquired sufficient strength, as soon as it comes in contact with the branch, pierces the bark, and plants itself in the alburnum, whence it sucks up its food, just as another plant would do from the soil.

CAROLINE.

With the advantage that its food is already prepared; it can therefore scarcely require leaves to convert the sap into cambium.

MRS. B.

I beg your pardon. The soil from which it feeds is the wood, not the bark ; it is therefore the rising, not the descending sap which it receives ; the mistletoe and the tree to which it adheres, may therefore be considered as the same individual plant. The parasite receives the sap after the same manner as the branches of the tree, and like them, requires leaves for its elaboration.

EMILY.

This junction is very analogous to a natural graft.

MRS. B.

On the contrary, it is quite the reverse. In a graft, it is the vessels of the liber which unite ; whilst the mistletoe strikes its little root through the bark into the wood, and the junction of the vessels takes place in the alburnum.

Is it not wonderful that so young and tender a root should be able not only to pierce the bark, but even to penetrate the wood?

MRS. B.

It is, indeed; but observe that it does not go deeper into the wood than the external layer, which being the last formed, is the most tender.

EMILY.

Then it cannot be so difficult to root out a mistletoe, as I have heard?

MRS. B.

You must recollect, that every year a new layer of wood grows over the root; so that without having, itself, penetrated further, it becomes annually buried deeper; and after some years' growth in so hard and compact a soil, that there is but little chance of being able to extract it, without wounding the branch beyond recovery; the only mode of effectually extirpating it, is to cut off the branch to which it is suspended; it is better to lose that, than to suffer the tree to be molested by so disagreeable a companion.

The mistletoe is more partial to some species of trees than to others; but the oak is the only one almost wholly exempt from its depredations.

CAROLINE.

I thought that the mistletoe attached itself to

the oak in preference to all other trees, and that
the Druids considered their union as sacred.

MRS. B.

It was probably owing to its so seldom attacking
this tree, that the Druids held it in such high vener-
ation when they found it there.   This tree has,
however, another enemy, of a very similar description, called the *Laurientius*, which confines its ra-
vages to this sovereign of the forest.

The *Cuscuta*, commonly called *Dodder*, is a parasite, which attacks lucerne, trefoil, and several of
the artificial grasses: it has neither cotyledon nor
leaves, consisting simply of a sort of filament or
stalk, which, after it has sprouted, falls and perishes,
when it finds no plant to which it can adhere; but
if it meets with any of the artificial grasses, it fastens
upon them, and feeds upon their juices.   The
seeds sometimes germinate in the soil, and sometimes on the artificial grass itself.   The mode of
destroying this noxious parasite, is either to burn
or to mow the artificial grass very frequently, in
order to prevent the seed of the Cuscuta from germinating; or else to change the course of cropping, and sow corn, for this parasite will not attack
grain, or any endogenous plant.   There are three
species of Cuscuta, one of which attaches itself exclusively to the vine: its filaments are as large as a
small packthread; fortunately, this last is very rare.

The *Orobanche* is a genus, one species of which
adheres to the roots of hemp, and destroys them
by devouring their juices.

Fungi form a very considerable class of false parasitical plants ; to this class belongs the *Erisiphe*, which attacks the leaves of plants : it first makes it appearance under the form of yellow spots, which afterwards turn black.   There are no less than forty different species of this parasite.

The Rhizoctonia is a species of fungus, which confines itself almost wholly to the roots of lucerne and saffron : this disease manifests itsfelf by the fading of the head of the plant, and the contagion soon spreads around it, in rays as from a centre. If one of the affected plants be pulled up, the roots will be found covered with the noxious filaments of this fungus : their effects on saffron is so baneful, that the malady it produces bears the name of *death ;* and the only way to prevent its spreading, is to bury the affected plants in a sort of cemetery, for it is necessary to surround them by a ditch; and in digging it, care must be taken to throw the earth inwards, to prevent the contagion from spreading.   There are three species of this destructive fungus, the brown, the carmine, and the white : the latter attacks fruit trees ; its filaments are free from tubercles, while those of the former are covered with them.

The class of internal fungi is very numerous, there being not less than three hundred species, each attaching itself to the plant which suits it. Some of them attack all the plants of the same family ; others confine themselves to those of the same species.   Two of those species of fungi belong to the rose-tree : they appear at first under the form of small yellow spots; these increase till they run into each other; their colour then changes to various tints of brown and red, tints which you must have observed the leaves of the rose-tree often assume, long before their natural decay.

CAROLINE.

This malady, far from disfiguring the plant, adds to its beauty; but who would ever have imagined these colours to have proceeded from a separate vegetation growing on the leaf?

MRS. B.

Smut is a fungus, under the form of a black powder, which lodges itself on the surface of the ears of corn, particularly of oats. But the most insidious enemy of grain, of the mushroom tribe, is called the Rot. It devours the seed, without making its appearance externally. When the corn is thrashed, the rotten seeds burst, and the disease is thus communicated to the rest of the corn; so that if sown, the rot will be propagated as well as the corn; to prevent which, corn that is at all affected with this disease should be soaked in a lime wash, which destroys the seed of the rot, without injuring that of the corn.

Mr. Benedict Prevost has found that washes of vitriol or verdigris, are still more efficacious.

EMILY.

But are they not pernicious to the grain, and even dangerous to those who employ them?

MRS. B.

It was at first apprehended to be so, but it is now well ascertained, that neither the labourer nor the grain suffer from this process : it is much used in France, and even arsenic has been tried with success for this purpose.

The Ergot is a disease peculiar to rye, which attacks the ovary of that plant; and bread made of rye thus affected is extremely unwholesome, frequently producing gangrene.

It would be endless to detail the various fungi which molest the vegetable kingdom ; we will conclude, therefore, with the *Rust,* which confines its depredations to the grasses.

It is time now to turn our attention to the last class of diseases, those resulting from age ; and here you must observe that a very essential difference exists between the animal and the vegetable creation. In the former, all the organs are developed at once : these after long use become indurated, obstructions take place, decay follows, and life thus often terminates from old age. But the economy of the vegetable kingdom is totally different : the organs of the plant, that is to say, the vessels which convey the juices, the leaves which elaborate them, the buds which produce flowers and fruit, are renewed every year ; they are always fresh, always young : how then can a plant decay from age ?

CAROLINE.

I should rather ask why all plants do not, like

annuals, die every year? for these organs which are renewed in the spring, perish in the autumn.

MRS. B.

Of all the organs which are annually renewed in perennial plants, the layer of wood and of bark alone survive in an active state of vegetation: the others may be considered as annuals, living but one season.

CAROLINE.

Then, when a tree dies of age, it is from the stem being worn out: every year the wood hardens by the pressure of the new layers which grow around it: its vessels must, in consequence, become obstructed, and less adapted to convey the fluids which are to pass through them: this bears a strong analogy to the decay and death of animals.

MRS. B.

True; but observe that if these vessels are no longer calculated to transmit the juices, the plant no longer requires them to execute this function: it is performed by the fresh layer of wood and of bark which are renewed every year: the old repose after their labours, but do not perish; age, therefore, does not necessarily entail death, as in the animal kingdom.

EMILY.

From what cause, then, do plants die? for, though some trees live to a great age, they all ultimately perish, as well as animals.

MRS. B.

They are certainly not destined to immortality; but their ceasing to exist seems to depend upon some accidental disease proving fatal, rather than upon any prescribed term of years assigned to them by Nature.

The malady which most commonly destroys plants is exhaustion, arising from their bearing, and ripening, too great a number of seeds: it is this which regularly, though not necessarily, occasions the death of annuals; for, if from any accidental circumstance the seeds are not matured, the plant retains sufficient vigour to live through another season. Perennials, which live several years, perish ultimately of the same disease.

CAROLINE.

And are there no means of diminishing the number of seeds of annuals, and by thus preventing exhaustion, of transforming them into perennials?

MRS. B.

This may be done by making the flower grow double: the additional number of petals are produced at the expense of the seed; but requiring much less nourishment, the plant is not exhausted.

EMILY.

But, if trees perish only by accidental death, some, at least, should escape; for accidents do not always occur.

MRS. B.

Not, perhaps, to a certainty in any given period; but in the long course of time, they never fail to happen; and the extreme inequality in the length of life, in trees of the same species, affords ground for believing that its duration depends upon accident.

EMILY.

But some kinds of trees are regularly much longer-lived than others: the oak, for instance, than the poplar; forest, than fruit-trees.

MRS. B.

Some plants are naturally much more robust than others, and therefore resist during a longer period accidental attacks. The oak, so vigorous and magnificent a tree, out of six seeds which it produces in every blossom, brings only one to maturity; and yet with how much less effort could the oak ripen clusters of acorns than an orchard tree the heavy load of fruit, under the weight of which its branches bend; and if any of them break, how great is the probability that decay will ensue: the enfeebled vessels of the wood, exhausted by the labour of conveying sap to so much fruit, are unable to resist the consequences of exposure to the weather; and, after a series of accidents of a similar nature during a course of years, the tree at last perishes. When, therefore, it is said that such a species of tree, usually lives such a number of years, the duration refers to the average of time in which it falls a sacrifice to accident. This average it is very difficult to ascertain.

But it is not only ripening seed, which eventually exhausts plants; all the various diseases we have enumerated tend to shorten their existence.

EMILY.

Yet those only which attack the wood or bark can prove dangerous to the life of the tree: injury to the other organs, can be of little consequence, since they naturally perish in the autumn.

MRS. B.

True; but observe that most of the diseases we have mentioned are of that nature; the parasites suck up the juices of the stem; the fungi which adhere to the stem and branches, those which coil round and strangle the roots, all eventually injure the wood.

There is in the island of Teneriffe a tree, the Dracæna Draco, of so remarkable a size, that it served to point out the limits of possession of different tribes when the island was first discovered: it has since been repeatedly visited by different travellers, and during several centuries past appears to remain unchanged: it may possibly be of so vigorous a nature as to have existed some thousand years.

CAROLINE.

And the extraordinary large tree in the Cape

Verde islands, Mrs. B., in which Mr. Adamson discovered an inscription buried under three hundred layers of wood, must have been of a very great age.

MRS. B.

From its dimensions and appearance, he calculated that it was probably about five thousand years old.

CAROLINE.

Even allowing for an error of a thousand years or two in his calculation, the tree would still be of a very venerable age.

And without going so far for an example, in Blenheim Park there are still in existence old trunks of trees, which are said to have shaded the retreat of the fair Rosamond; and they are supposed to be now not less than a thousand years of age.

CONVERSATION XXVIII.

ON THE CULTIVATION OF TREES.

MRS. B.

NATURE has divided the surface of the earth into meadows and forests: in some parts of the globe, these are so happily blended as to form the most beautiful variety of prospect; but in general, where the hand of man has not interfered, they are divided into immense masses of wood and pasture, which render the appearance of the country monotonous and melancholy.

EMILY.

I should have thought that, in the course of a series of years, these different species of vegetation would have intermixed, so that the seeds of the forest trees would have sown themselves and grown up amongst the grass, while the latter, on the other hand, would have spread amongst the trees and gained ground upon the forest.

MRS. B.

On the contrary, these two species of vegetation reciprocally interfere with each other, so as to prevent either from encroaching on their established limits; for grass will not grow under the impenetrable shade of a forest, nor will the seeds of trees germinate in those thick and rich wild pastures called Steppes, where the grass rises to six or eight feet in height.

In tropical climates, forests are composed of a much greater diversity of trees than they are in our less genial latitudes; and the more you travel northward, or the greater the elevation of the land, the more homogeneous the woods become.

CAROLINE.

I have observed this, both in travelling in Scotland and in ascending the mountains of Switzerland. The walnut, the oak, and the birch successively disappear, and the summits are almost always crowned with firs.

MRS. B.

It is remarkable, that under the same latitude, America can boast a much greater variety of trees than Europe: we possess but thirty-four species, while she has no less than one hundred and twenty. It is to be hoped, at least, that we shall be able to increase our stock from so well furnished a market.

EMILY.

America being a more recently settled country, and less populous, can afford to raise wood in a better soil, whilst we, in Europe, are so restricted for space, that all our good soil is set apart for grain, and we plant wood only where nothing more valuable will grow.

MRS. B.

That is necessarily the case in all highly civilised and thickly populated countries; corn being a more valuable produce than timber, will obtain the preference where the soil is adapted to it.

Our natural forests, in such land as we allow them to occupy, consist of little more than the oak, the ash, the beech, the birch, and, in elevated situations, the fir.

Forests are divided by botanists into tolerant and intolerant: the former admits of trees of another species growing amongst them; the latter exclude all but their own.

A forest of oaks is of the former description; underwood of various descriptions growing beneath it; whilst beeches and firs do not allow this privilege to the inferior plants, and are hence denominated intolerant.

The thirty-four species of European forest-trees are divided into four classes.

1st Class. Trees with hard wood: this class comprises three species of oak.

Long-stalked oak.	Chesnut.	Hornbeam.
Stalkless oak.	Elm.	Pear.
Tauzin oak.	Ash.	Apple.
	Sycamore.	

2d Class. Trees with soft wood.

| Lime. | Poplar. | Willow. |

3d Class. Trees with resinous wood.

Pine.   Fir.   Larch.

4th Class. Evergreens, not resinous.

The Evergreen Oak, of the south of Europe.

There are two modes of felling forests; which the French call *en Jardinier*, or *Taille réglé*, and for which we have no equivalent terms in English. In the former, you successively cut down the large trees as they grow up to the size of timber; in the latter, the whole of the forest is felled at once.

CAROLINE.

The former must surely be the best mode; for it seems mere waste to cut down the young trees before they are large enough to be of use.

MRS. B.

It is difficult to fell the large trees without injuring the small ones. They are deprived of the shade and shelter of the large trees, and their roots are often deranged and their branches broken by the fall of their protectors. When forests are felled completely, it is done at regular periods, which are determined either by the nature of the wood or the purpose for which it is intended. For the ordinary consumption of fuel, it is usually cut down every twenty years. When the trees have attained a sufficient size for fire-wood, and in countries where wood is the only fuel, this is the principal object in view. This is generally the case in most parts of the world. Our island, where coal is so commonly burnt, forms an exception. We devote much less time to planting, because we derive our fuel from the bowels, rather than from the surface, of the earth ; and our woods are raised chiefly to produce timber for building : but the *taille réglé* is generally adopted on the Continent. It admits, however, of some modification : instead of cutting down the large trees, and leaving the young ones to grow up, the young trees are cut down generally about the age of twenty years ; with the exception of the finest plants, which are reserved for the next periodical felling. These trees are called standards or standers. The young trees are cut up into faggots for burning, and into props to support vines : their stumps quickly send forth new shoots, which at the end of another twenty years are fit to be cut down for the same purpose. The greater number of the standards are then felled, having acquired dimensions which enable them to be cut into logs for fire-wood. The standards which escape the second felling, in France, assume the name of *sur taillis* : if reserved a third time, they are called *sur écorce ;* and should they be so fortunate as to survive the fourth felling, they become timber.

EMILY.

I really quite tremble for the reserved trees

every time the wood-cutter enters the forest; it is well they are not endowed with a consciousness of the risk they run. And at what age are timber-trees felled?

Their length of life will be more likely to excite your envy than your compassion. Oaks and beeches are not considered as ripe for timber, until they have attained the age of 120 or 130 years. After that period, there is a greater chance of their deteriorating, than of their improving.

CAROLINE.

Wood for burning, then, cut down in the successive fellings is from twenty to nearly fifty years of age?

MRS. B.

Beech, when not reserved for timber, is not suffered to live beyond thirty years; because, after that age, young shoots will no longer sprout from the old stumps.

Resinous trees do not shoot out afresh after felling; woods of firs must therefore be cut down altogether, and resown; or the forest may be felled in alternate stripes, which is attended with this advantage, that the stripes left growing shelter the young plants which shoot in those that have been felled. When the firs are situated on the declivity of a mountain, as it very frequently happens, the wood-cutters must begin their operations from below, in order to be able to carry away the trees with greater facility.

The proper season for felling forests, is from the middle of November to the middle of April; and the instrument best adapted for that purpose is a sharp axe, which should be used as near the ground as possible; the buds of the old stumps shooting much more readily when an axe is used than a saw.

EMILY.

And pray what species of trees do you consider as making the best fuel?

MRS. B.

That which is heaviest: the weight indicates the quantity of carbon it contains; and you may recollect that, in the combustion of wood, it is the carbon which gives out most heat.

CAROLINE.

Yet I should prefer the wood which produces most flame. Flame is so cheerful, that it appears, perhaps, to give out more heat than it really does.

MRS. B.

The light which accompanies it, in a great measure, atones for its deficiency of intensity; but flame must be considered as a species of luxury, in which those only can indulge who do not aim at economy in fuel. The greatest quantity of heat is given out when the wood burns red, without flame; consequently, the wood which has the fewest volatile parts producing flame, and the greatest quantity of carbon, producing red heat, is the most valuable for fuel. Here is a list of the various proportions of carbon contained in equal quantities of wood of different species : —

	Oz. of Carbon, the Cubic Foot.		Oz. of Carbon, the Cubic Foot.
Black fir - - - 86		Beech - - 64	
Red fir - - 84		Pear - - 54	
Evergreen oak - 69		Birch - - - —	
Box - - - 68		Willow - - 27	
Oak - - 60		Poplar - - - —	

I have already mentioned the danger incurred by stripping trees of their bark, in order to harden the wood, by forcing the cambium to descend through it. This mode is, however, sometimes attended with success, provided that it be performed only the spring previous to the trees being felled ; and that the naked tree be charred or slightly burnt, as a substitute for the covering of which it has been deprived, and a preservative against the inclemency of the weather.

CAROLINE.

That is to say, that the wood is burnt to save it from suffering from wet ! I should really think the remedy worse than the disease. In travelling in Italy, I recollect seeing such miserable flayed and blackened trees, looking as if they had been put to various species of torture before

L 3

the executioner came with his axe to strike the final blow.

EMILY.

They reminded me rather of poor naked savages, smeared and tattooed as a substitute for their deficiency of clothing.

MRS. B.

Why should you not compare them to sheep shorn of their fleeces in the spring? — the bark of the tree is no less useful in the arts than the fleece of wool: you recollect that it contains the astringent principle called tannin, so essential in the preparation of leather: it is oak bark which is principally used for this purpose, as it contains the greatest quantity of tannin.

Let us now consider the cultivation of single trees.

EMILY.

Such as form the ornament of parks and pleasure grounds; and those which, dispersed throughout the country, produce such beautiful scenery in England.

MRS. B.

There is certainly no country which can boast such a natural, and picturesque arrangement of trees. On the Continent, single trees are generally planted in rows : in some districts they may be considered as a supplement to forests. Almost all the trees in Belgium, for instance, grow in hedgerows, or in avenues on the side of high roads.

What species of trees should you think best calculated for the latter purpose?

CAROLINE.

Evergreens would not be suitable; at least in
our northern climates, because the road requires
exposure to the sun and wind during winter, and
the passenger requires no shade in that season.

EMILY.

They should be trees which afford sufficient
shade in the summer, but whose foliage is not so
thick as to prevent the road from drying, after
heavy showers.

MRS. B.

For this reason the horse-chesnut is not adapted
to such a situation; for, though it would afford excellent shelter to passengers during a shower, it
would render the road damp. Trees with very
wide-spreading roots are also objectionable, as they
encroach on the adjacent culture: on this account
the acacia is excluded.

EMILY.

But in England the road is almost always separated from the contiguous fields by a ditch, so
that the roots could not well interfere with their
produce.

MRS. B.

The roads in England are seldom bordered
with trees, except occasionally in hedge-rows:
our climate being too damp to admit of such an

ornament, while in the southern parts of the Continent trees are almost a necessary accompaniment to roads, on account of the shade they afford.

CAROLINE.

I think fruit-trees, and such as have sweet-smelling blossoms, should be planted for the gratification of the passengers.

MRS. B.

I am afraid that the kindness of your intention would be frustrated; for as these trees are not the property of the public, it would be only leading the passenger into temptation, and exposing the tree to danger; the fruit would be unlawfully gathered, and eaten, in all probability, before it was ripe; and the tree would suffer from the pulling and breaking of its branches. When fruit-trees grow on the high road, the proprietors are often obliged to fence the stems with briars and brambles, to prevent their being climbed, and to lop the lower branches, in order that they may be above the reach of the passengers.

The elm is one of the trees best adapted for the high road: it may be transplanted without injury, after it has attained such a growth as to enable it to resist the attacks of cattle, an essential point for trees so much exposed: it is robust in its nature, of long duration, and affords a light and pleasant shade: its roots are superficial, and yet not spreading, and it bears neither flowers nor fruit which can tempt the passenger.

The plane, a tree very common on the Continent, is also well adapted to roads. It comes into leaf very late, so that the roads have full time to dry in spring. The oak is so robust and durable a tree, that it would be excellent for this purpose, could it be transplanted sufficiently large to preserve it from accidental injury; but it suffers from transplanting unless very young. The best mode of rearing oaks for avenues is to plant them in hedges: the bramble, or other shrubs of which the hedge is composed, afford them shelter and defence, until they are of an age to resist accidental injury; the hedge may then be cut down at pleasure. This has also the advantage of forcing the roots of the oak to descend; for the roots of the hedge, being more superficial, consume the nourishment near the surface of the soil, and compel those of the oak to seek it in a lower region.

The birch is well calculated for roads, if the soil be sandy; it thrives in a cold climate and in elevated situations, and the lightness of its foliage is an additional advantage in such temperatures.

The sycamore is a beautiful tree for avenues. The hornbeam is objectionable only on account of the slowness of its growth. The aspen, the ash, and the poplar, are well adapted to a moist soil, as they help to drain it. An avenue of poplars is not picturesque, it is true, but it affords almost as much shelter from the wind as a wall, and in some situations this is very desirable.

EMILY.

A few poplars, interspersed with other trees, form, I think, beautiful groups; but an avenue of poplars is associated with the idea of marshy ground, and from its formality is extremely ugly.

MRS. B.

In planting trees by the road-side, the holes should be made both deep and wide; for the ground not being cultivated is hard and compact, and the young roots would be unable to penetrate it, were it not prepared and lightened by the pickaxe or the spade.

The young trees should never be headed or lopped; it thickens their foliage, but destroys the natural character of the tree. Some of their lateral branches may be slightly pruned; for as the branches in general correspond with the roots, the more erect the former grow, the more the roots will descend into the soil.

CAROLINE.

The beauty of the greater part of the trees on the Continent, is spoiled by the merciless mode they have of heading them when young, in order to make them grow thick and bushy.

MRS. B.

In transplanting young trees, the more they are lopped, the more certainty there is of their living; and nurserymen who usually supply them, and warrant their taking root, make no scruple to amputate both head and branches.

CAROLINE.

The life of the plant may be thus secured, but it no longer deserves the name of a tree; it is a stake, or a pole; which, though it may throw out branches, will never have the free, natural character of its species. Is it not, therefore, far preferable to run some trifling risk of losing them, rather than mutilate them in so barbarous a manner?

MRS. B.

I perfectly agree with you: besides, the risk is so trifling. In transplanting trees into the botanical garden at Geneva, the branches are never lopped, and only about five per cent. die; yet the chance of their perishing must greatly exceed that of common transplantations, as the trees come from foreign climates, and are placed in a soil and temperature more or less unsuited to them; besides which, they have undergone the confinement of packing, and the fatigues of a long journey. It must, however, be acknowledged, that the art of packing and conveying plants is highly improved; for M. De Candolle frequently receives plants, not only in leaf, but in full blossom.

If, however, gardeners will persevere in this system of lopping, they should at least do it with moderation and judgment.

There is, however, some apology for nursery
gardeners: the trees, in their grounds, are so thickly
planted, that they cannot be taken up without injury to their own roots, or to those of their neighbours; and if the roots be cut, is it not necessary
also to lop the branches, for mutilated roots can ill
supply the whole of the branches with nourishment?

Your observation is very just; but young trees,
raised with a view to transplantation, should never
be allowed to grow so thickly, as to interfere
with each other; it is the duty of a nurseryman
to transplant them in his own grounds, in order
to give them space, if he has not a market for them
elsewhere.

When a very large tree is to be transplanted,
it is advisable to do it in the heart of winter,
during a frost; a trench should be dug around,
and, as far as attainable, below the stem of the
tree, and be filled with water, if the rain does
not sufficiently perform that office. When the
water is frozen, the tree will be inclosed, as it were,
in a vase of ice, and may be taken up with the
clod of earth attached to it; and, contained in the
icy vase, it may then be conveyed to the place of
its destination, almost without being sensible of its
change of situation. This is, however, a very
expensive operation, as it requires a considerable
mechanical force to accomplish it   it can be done,

also, only where the frost is severe and of long duration; for if the vase be melted or broken before the tree is placed in its new situation, the whole fails.

Sir Henry Stewart of Allanton has, within a few years, introduced a mode of transplanting large trees, which appears to have been attended with great success. It is precisely the reverse of that I have mentioned, yet founded on the same principle of guarding the roots from injury: with this view, instead of carefully covering up the roots, he lays them bare, but he separates the earth from them with such extreme precaution, that not even the smallest fibres are injured: this is done by labourers, whom he calls pickmen; because their business is to clear the roots from the earth by means of a small instrument adapted to the purpose, or with their fingers; a ball of earth is left close to the stem with the sward upon it. An engine is then brought up to the tree, consisting of a strong pole mounted upon two high wheels; the pole is strongly secured to the tree, while both are in a vertical position, they are then brought down to a horizontal one, by the pole acting as a lever; and by its descent, the few central roots, which the pickmen could not reach, are rent from the ground. The tree is so laid on the machine as to balance the roots against the branches, and one or two men are placed aloft among the branches of the tree, where they shift their places like movable ballast, as occasion may require. Both roots and branches are carefully tied up. The pit for receiving the tree, which should be prepared a twelvemonth before, is now opened, and the tree set in the earth as shallow as possible. The roots are then loosened from their bandages, and divided into the tiers, or ranks, in which they grow from the stem; the lowest of these tiers is then arranged, as nearly as possible, in the manner in which it lay originally, each root with its rootlets and fibres being imbedded in the soil with the utmost precaution, the earth being carefully worked in by the hand and the aid of a small rammer: additional earth is then gradually sifted in, and gently kneaded down, till it forms a layer, in which the second tier of roots is extended in the same manner as the lower tier, and so on till the whole are covered with earth. This attention to incorporate each fibre of the roots with the soil not only answers the purpose of inducing the roots to recommence their function of absorbing sap, but also serves to fix and secure them firmly in the soil, and renders stakes, ropes, and other means of adventitious support unnecessary.

CAROLINE.

And by your account this does not appear to be a very expensive process.

MRS. B.

No; independently of the engine, which is very simple, it is estimated that trees from twenty-six to thirty-five feet high, may be moved half a mile, at the expense of from ten to thirteen shillings. But the experiments have always been made with healthy trees, whose roots and branches have had ample space for growth ; not tall emaciated plants torn from the interior of forests, with stinted roots and branches, and so little vigour of vegetation, that their bark would not be either of sufficient thickness or hardness, to shelter the stem from the rude blast, nor the roots of sufficient strength or extent, to fix it firmly in the soil. Sir Henry Stewart, therefore, particularly recommends transplanting trees which have been freely exposed to the advantages of light and air; and should they, by such exposure, have suffered in their growth on the weather side, he advises, in transplanting them, to reverse the aspect, in order to shelter the weak side of the tree, and expose the luxuriant one to the severity of the wind.

EMILY.

By this means, then, large trees may be transplanted without either cutting the roots or lopping the branches?

MRS. B.

By adopting all the precautions I have mentioned, it appears that scarcely a tree failed. Sir Henry is, no doubt, perfectly correct in not cutting, even the little tassels of rootlets which grow at the extremities of the roots, provided the operation of transplanting be performed with so much caution that these suffer no injury; but if the spongioles be crushed, or the fibres any way mutilated, it is better to amputate the extremities, which will shoot afresh, more quickly than the rootlets would recover of their wounds.

It is wrong to plant in wet weather; for, though watering is required after planting, the hole in which the tree is placed must not be filled with mud: it would greatly endanger the roots.

In such wet countries as Holland, they are often obliged to bury faggots beneath the soil intended for planting, in order to increase the filtration of the water.

From the cultivation of trees we shall proceed to that of hedges: these are destined either for shelter or defence. In former times there was a third description of hedges, designed for ornament; but our landscape-gardeners have entirely exploded the grotesque figures, cut out in box and yew, which excited the admiration of our forefathers.

That district in the west of France, called the Boccage, derives its name from the high and bushy hedges with which it abounds, and which are designed to afford shelter from the stormy winds of the Atlantic. There are but few trees in those parts; but the hedges, being from eight to ten feet in height, are sufficient to protect 'the crops from the boisterous sea-breezes, and they thence bear the name of *brise vent*.

CAROLINE.

In England, our hedges are calculated more for defence; but the trees, with which they are interspersed, serve also the purpose of shelter.

MRS. B.

Our climate is unfortunately so damp, that exposure to the sun and air is rather an advantage than otherwise.

Hedges for defence answer the double purpose, of enclosing cattle in their pastures, and excluding those which might trespass on it.

EMILY.

Does it not also afford a security against thieves?

MRS. B.

It has been so considered, but, I am inclined to think, erroneously. A thief may lie concealed, and lead away a sheep or a cow at night, under cover of a hedge, without being discovered; whilst there is scarcely any night so dark, that he might not be perceived on an open plain.

It is objected to hedges, that they occasion a waste of ground: when necessary, therefore, they should be made to occupy as little space as possible, and be thickened, by crossing and engrafting the branches on each other, rather than by planting a double row. An external ditch is liable to the same objection; but it has the double advantage of serving as a defence to the hedge, and of

raising a bank, which gives additional elevation to
the hedge when planted on it.   When the shoots
are two years old, they may be crossed and fastened
by a worsted thread, and they will engraft of themselves; for the friction of the ligature will wound
the young bark sufficiently to expose the cortical
vessels, and enable them to unite with each other.

EMILY.

The plants have, then, a double source of life;
and, if one of the stems should perish, its branches
would be fed by those on which it is grafted.

MRS. B.

Yes; and the dead stem may be cut away without injuring the hedge.   By this system of crossing and grafting the branches, the hedge becomes
so thick as to be absolutely impassable.   Great
attention should be paid, not to plant hedges of
shrubs which grow thin at the base, or have spreading roots.   The hawthorn or quickset is decidedly
the plant best adapted for hedges ; its shoots
branch out in such a variety of directions, and
cross and intersect each other so frequently, as to
render all ligatures for that purpose unnecessary.

The Paliurus aculeatus succeeds well in dry
soils.   It is armed with two species of thorn, one
of which is straight, the other curved : so that the
animal that would trespass, if it can avoid the
straight thorns, on entering the hedge, has very

little chance of escaping the crooked ones in passing through it.

The barberry is well adapted for hedges, having three thorns issuing from the same point. The Ilex is furnished with thorns at the extremity of its leaves. The Lentiscus (Pistachia Lentiscus), and the Cockspur Hawthorn (Cratægus crusgalli), are shrubs which admit of planting in hedge-rows; but their cultivation does not extend further northward than the southern parts of Europe.

CAROLINE.

I begin to think we have been confined long enough by these hedges; and I am impatient to break through them, to get into the orchard, and examine the fruit-trees, which are of a much more interesting nature.

MRS. B.

I was just going to direct your attention to them.

You will be surprised to hear that, of one hundred and twenty families of fruit-trees, known in Europe, we cultivate only seventeen; and by far the greater part of these have been brought from the other quarters of the world. The apple and the pear, some few cherries, and the raspberry and strawberry, are alone indigenous in Europe.

CAROLINE.

Alas! what a poor figure our quarter of the globe makes in the vegetable kingdom!

EMILY.

We have the greater merit in having enriched it with such a number of foreign plants.

MRS. B.

True. These seventeen families give us thirty-four genera, sixty-eight species, and, finally, about two thousand varieties of fruit-trees : a number which is multiplying every day, from the increased facility of intercourse with foreign countries, and the improved mode of conveying plants, united to the general progress of science.

CAROLINE.

From what countries do we derive our choicest fruit-trees?

MRS. B.

Chiefly from the East. Africa is but very imperfectly cultivated; and America, though so distinguished for its forest-trees, appears to be very scantily supplied with fruit-trees. The Opuntia, the Diospyros, and a few others, are the only fruit-trees that have been brought to Europe from the northern parts of the New World.

The orange and citron we derive from Japan; the pomegranate from Africa.

New Holland, which contains not less than three or four thousand different plants, has but three or four species of fleshy fruit-trees, and the fruit of these is small and insipid.

In some fruits, we distinguish those in which the fleshy part is attached to the nut or kernel, as the plum and the peach, and those which are separated from it, as the apricot. Peaches, plums, apples, and pears, are of the family of *Rosaceæ*.

CAROLINE.

This family is, then, equally celebrated for the beauty of its flowers and the excellence of its fruits.

MRS. B.

There are two species of peach, both of which we derive from Persia: one of them, having a smooth skin, we distinguish by the name of Nectarine. Each of these species has two varieties, in one of which the pulp adheres to the stone, in the other it is separate from it.

The other members of this family are the almond, the apricot which comes from Armenia, and the cherry, of which there are five species. There are besides, of this family, the plum, the strawberry, the rose, the service-tree, and the medlar.

The orange forms a family of its own, bearing its name *Aurantiaceæ*, and includes the lemon, the citron, and the pample, or mousse.

The sweet orange and the bitter were formerly supposed to be of the same species, and the sweet was often grafted on the bitter orange; but this is an error: they are of different species, — and the sweet orange does not require grafting.

14

There are no less than twelve known species of
walnut-trees; one of which we derive from Syria,
and the eleven others from America. We cultivate the first for its fruit, but the latter produce
the finest timber.

CONVERSATION XXIX.

ON THE CULTIVATION OF PLANTS WHICH PRO-DUCE FERMENTED LIQUORS.

MRS. B.

THERE exists in all vegetables, though in very different proportions, a saccharine substance from which sugar is obtained; and this substance is susceptible of being converted into alcohol, or spirit of wine. For this purpose it is not necessary to resort to the laboratory of the chemist: when placed under favourable circumstances, the transformation takes place spontaneously by a process called *fermentation.*

EMILY.

It is a process with which we are already tolerably well acquainted, as you explained the different fermentations to us in our Conversations on Chemistry.

MRS. B.

You will then recollect that the juice of all fruits when expressed, will (like that of the grape) ferment; and that during this process a general disorganisation of the parts take place, and a new arrangement is established, in consequence of which the sugar or saccharine matter contained in the liquor will be converted into spirit. But fermentation is not confined to the juice of fruits: spirit may be obtained from any part of a plant containing the saccharine principle; thus the sap of the palm-tree, when fermented, produces palm wine.

CAROLINE.

It is to be regretted that we have no trees whose sap can be fermented: it would be so much more easily obtained than fruit.

MRS. B.

The sap of the birch is sometimes fermented. But you may recollect that the vinous fermentation is frequently followed by another of a very different nature, called the acetous fermentation, which reduces the wine or spirit to vinegar: this occurs in some measure with the fermented sap of the birch; it becomes slightly acid, and may therefore be considered rather as a refreshing than a spirituous beverage. All sap would yield spirit; but, independently of its susceptibility of turning acid, the liquor would, in general, be insipid. The excellence of wine is not confined to the spirit it contains, but to its aromatic flavour; and this is produced by the fermentation of fruit. If spirit of wine alone be required, it may be obtained from potatoes or any other vegetable, however insipid. Brandy and common spirits are, in England, usually distilled from fermented grain : gin has more flavour, as the juniper berries are distilled with the spirit.

Yet grain does not appear to contain any sugar ?

Though grain is not sweet to the taste, it contains the elements which produce sugar, and the mode of developing this substance, is to make the grain begin to germinate. For this purpose, barley is moistened and exposed to a certain elevation of temperature which stimulates germination; the saccharine principle is thus produced, and the grain becomes sweet: the germination is then suddenly stopped by drying the barley in a kiln or a heated oven; in this state it is called malt. When mixed with water, the liquor is so sweet as to have obtained the name of sweet-wort, and its fermentation produces beer; but this would be a very insipid beverage, were not hops added previous to the fermentation, to give it the flavour and astringent quality found in fruits.

The fermentation of apples produces cider. There are three species of apples, the sweet, the sharp, and the acid. The two former, fermented together, produce excellent cider: the sweet apple supplies the spirit; the sharp, the astringent prin-

ciple; but the sour apple is not fit for fermentation. In order to make good cider, it is not
only essential to choose the races of apples, but
they must be gathered with care to avoid being
bruised; they should then be collected into heaps,
in which state they ripen and exude moisture:
they must next be crushed and reduced to a
pulp, and $\frac{1}{20}$ of water added; the mass is then
pressed to obtain the juice, which ferments spontaneously, and produces cider.

EMILY.

Perry is, I believe, obtained from pears in a
similar manner ?

MRS. B.

Precisely. But it is to the vine that we are indebted for the most precious of our fermented
liquors. This plant is of the family called *Sarmentaceæ*. It bears alternately clusters of grapes, and
of leaves, opposed to each other on the stem. The
vine derives its origin from the countries situated
between Persia and India: it was brought by the
Phœnicians to Greece, and thence conveyed by the
Phocians to a colony they had formed in that part
of Gaul, where Marseilles is now situated.

EMILY.

The vine is a plant of such interest to society,
that its history can be traced with more accuracy
than that of most other plants.

MRS. B.

Hence governments have interfered more with the culture of the vine, than with that of any other plant. Numa Pompilius first introduced it at Rome. The Emperor Domitian ordered all the vineyards to be rooted up. Charlemagne protected the culture of the vine; whilst Charles IX. discouraged it. His successor, Henry IV., re-established it, and ever since it has flourished in France.

CAROLINE.

It appears, then, that cruel and tyrannical sovereigns forbad the culture of the vine, whilst the humane and enlightened ones encouraged it; and yet the former could have been influenced only by its moral effect on their subjects, for it was evidently prejudicial to the interests of the country, to destroy so valuable a branch of commerce.

MRS. B.

Commercial interest was very imperfectly understood in ancient times, especially by unenlightened sovereigns; these, therefore, considered only the prejudicial effects of the vine in producing intoxication, whilst the better informed not only esteemed it as a source of wealth, but of health and comfort to those who enjoyed it without excess,— and this latter class is certainly by far the most numerous.

EMILY.

I have heard it observed, that there is less in-

M 2

toxication in wine countries than in the more northern districts, which do not admit of the growth of the vine.

MRS. B.

In England, for instance, it is cheaper to drink spirits than wine, or even than strong beer; and as alcohol is the intoxicating principle, these distilled liquors have a more intemperate tendency. The culture of the vine, since its introduction into Europe, in extending northwards, has spread itself more to the east than to the west, because the eastern part of this continent is hotter in summer than the western, under the same latitude. Now, the vine derives more advantage from the heat of summer than it suffers from the cold of winter: in the latter season it does not vegetate, so that it requires only the degree of temperature necessary to escape freezing, while heat in summer is absolutely requisite to ripen the grapes; and you have seen that the vine succeeds much better in Switzerland than in England, because, though the winters in the former are generally colder, the summers are hotter.

CAROLINE.

We read in history, of vineyards growing, and wine being formerly made in England. Do you suppose that the climate was then warmer than it is now?

MRS. B.

No; but the palate of our ancestors was probably not so delicate as that of their descendants. The same has been affirmed of Brittany and Normandy, provinces in which vineyards are now unknown, and where the vine is cultivated, as in England, trained against walls in a favourable aspect; and even then the grapes ripen but imperfectly. If wine was really ever made in those countries, it must have been a beverage very analagous to vinegar; but it is very possible that such wine was once produced: for the fact is ascertained, that in proportion as the means of transport has increased, the extent of country, in which the vine is cultivated, has diminished.

EMILY.

I should have imagined that the increase of high roads, canals, and shipping, would, by diminishing the expense of conveyance, lower the price of wine, and thus render it more attainable to the northern countries, where it is not grown.

MRS. B.

Your argument is perfectly just: the increased facility of conveyance augments the demand for, and, consequently, the production of, wine; but that does not prevent its restricting the extent of latitude in which the vine is cultivated. When wine could be conveyed from the south of France to Brittany and Normandy, of a much higher flavour and better quality, than that which was produced in those provinces, and with but little additional

M 3

expense, the Bretons and Normans gradually converted their vineyards into corn and pasture, and exchanged their grain and cattle for the juice of the grape.

CAROLINE.

Whilst the increased demand for wine, must have induced the southern districts to convert their pasture and corn fields into vineyards. The same reasoning will hold good with regard to England; and wine must have been conveyed across the Channel, to the utter destruction of the English vineyards. You see, Mrs. B., that I have not forgotten your lessons of political economy.

MRS. B.

I am glad to hear you remember them so well: the cultivation of vineyards at present extends from 29° to 50° of latitude, as far south as Shiraz, in Persia; as far north as Cologne, on the Rhine.

EMILY.

Pray, does not the vine grow naturally in America?

MRS. B.

It does; but it is of a different species; and grows only wild; the vine which is cultivated, is brought from Europe; but its introduction has not hitherto been attended with complete success.

EMILY.

I am surprised at that, as the islands of the Atlantic, Madeira, and the Canaries are so celebrated for their wine.

MRS. B.

On the continent of America, all the grapes in the same cluster frequently do not ripen at the same time; so that, when gathered, some are decaying, whilst others are not yet come to maturity: and this circumstance, which is not yet accounted for, prevents the wine from being of a good quality.

It is at the Cape of Good Hope that the vine has made the most remarkable progress, and particularly since England has been in possession of that colony. Whilst it belonged to the Dutch, it produced only a small quantity of rich Cape wine; but now a variety of different vines are cultivated there with great success, and the Cape Madeira will, perhaps, ultimately rival that of the Atlantic island.

The height at which the vine can be cultivated, from the level of the sea, is four hundred fathoms.

EMILY.

But that must vary according to the latitude?

MRS. B.

No doubt; this is the elevation of the most northern limits of the cultivation of the vine in

France.   There are many circumstances to be attended to, in the culture of a plant of so much importance as the vine.   In the first place, the nature
of the plant: the varieties are innumerable; there
are no less than six hundred in the botanical garden of Geneva, the fruit differing either in colour,
form, flavour, consistence, &c. : the degree of flavour, of firmness and compactness of the fruit,
is, in general, proportioned to the heat of the climate.   The flavour of the muscat grape is, however, richer than that of the common grape in any
climate.

Every flower of the vine contains five seeds,
two or three of which often fail.

CAROLINE.

The soil must labour hard to ripen so many
seeds?

MRS. B.

No cultivation requires greater care to repair
the exhaustion which it undergoes, and attention
to prevent weeds from engrossing any portion of
that food which is so much in request.   Yet a
great deal of manure should not be used, for it
injures the quality of the fruit, though it increases
the quantity.

The grapes should be neither very close, nor
very distant from each other in the cluster, but so
far apart as to leave sufficient space for each grape
to attain its full growth.   For this purpose, the
grapes of Fontainebleau, when young, are thinned

by the scissors.   But these grapes are cultivated exclusively for eating, and sold at a price which repays such an expensive mode of culture.

There is also great diversity in the degree of precocity or tardiness of this plant.   When it shoots early, there is danger of its suffering from the frost in spring; if late, it may not have time to ripen its fruit in autumn.   Care, therefore, should be taken to choose the medium, especially in cold climates.

Old plants produce the finest fruit, but in the smallest quantity.   It does not, therefore, answer, to continue to cultivate the same plants, above a certain number of years.

CAROLINE.

So that they are not allowed time to meet in the course of nature with their accidental death?

MRS. B.

Not often.   The influence of climate on the vine is very considerable.   The greater the degree of heat, the more sweetness is developed in the fruit, the greater is the quantity of alcohol produced by fermentation, and the astringent principle is proportionally diminished: but this may be carried too far; a certain admixture of the astringent principle is both wholesome and palatable.   The grapes of Fontainebleau will not produce good wine, from not possessing a sufficiency

M 5

of this principle; and, accordingly, we find that
the wines in highest estimation are not those produced in the hottest climates, but in countries
situated between 30° and 45° of latitude. The
aspect most favourable must be determined by the
locality of the situation and the latitude. The
vines of Epernay, which produce the finest cham-
pagne, have a northern aspect; those situated on
the two opposite banks of the Rhone, in the
neighbourhood of Avignon, yield equally good
wine: but, in colder climates, the more vineyards
are exposed to the south the better they thrive.

It is rather singular, that fine grapes may be
produced in almost every kind of soil, provided
the vine be of a nature to suit it. The vineyards
of Bourdeaux are planted in a gravelly soil, and
hence bear the name of *Vin de Grave ;* those of
Burgundy in calcareous clay; hermitage grows in
granite; and Lacryma Christi is raised in the vol-
canic soil of Mount Vesuvius. The vineyards of
Switzerland grow in stiff, compact, calcareous
earth.

In order to determine upon the mode of culture, the question must first be ascertained whether
it be grapes of the finest quality or in greatest
quantity that are required. In hot countries, the
former are most in demand; in cold countries, the
latter is principally aimed at : for in districts which
form the limits of the cultivation of the vine, it is
desirable to produce a large quantity of wine, though
it be of inferior quality, for the beverage of the

common people, who cannot afford to pay the conveyance of wines from more favourable climates.

An argillaceous soil produces but indifferent grapes, even in a favourable climate. Under such circumstances, therefore, quantity rather than quality is aimed at, in order to obtain spirit for brandy; for in wine countries brandy is distilled from wine rather than from grain. For this purpose, the plants, instead of being kept low, as you have seen in France and Switzerland, are allowed to grow to a great length, and are suspended in garlands from one tree to another.

CAROLINE.

This mode of cultivation is adopted in Italy for the production of grapes for wine, and is most beautiful in appearance.

MRS. B.

But the fruit is not of so fine a quality; and, consequently, the wine is not so good. In the south of France, as well as in Italy, vines are often cultivated without being propped, and the branches are suffered to grow six or eight feet in length.

During the wars of the Revolution, the French having destroyed all the props of the vineyards in the valley of the Rake, on the banks of the Rhine, the peasantry were obliged to let the vines grow without support; when, instead of being deteriorated, they found the fruit so much improved, that they have ever since continued the same system.

EMILY.

Then I conclude that they did not allow their vines to shoot out to a great extent.

MRS. B.

No doubt; or the fruit would have been impoverished, instead of being improved. The use of props in vineyards is, perhaps, carried to the extreme. M. De Candolle suggests the experiment of fastening four plants to one prop, placed in the centre. In doing this, the branches would be curved towards the prop, and the descent of the cambium retarded.

CAROLINE.

We have seen vineyards in some parts of Italy trained on a horizontal trellis-work, the grapes being suspended beneath the verdant roof. In other places, the vines are trained over trees, which are planted merely to afford them support, and they derive a little shelter from the leaves, which grow on the few branches which are not lopped.

MRS. B.

These various modes of training vines, though they may be used in hot climates with less injury to the fruit, never fail, more or less, to be prejudicial to it; and though the climate of Italy is generally hotter than that of France, the latter is

celebrated in all parts of the world, for the excellence of its wines, while those of the former are
scarcely ever exported.

In hot climates, the grapes are sweet, contain less acid and astringent principle, and the
fermentation is less complete, the proportions not
being so well adjusted as in France, and other
countries of a more moderate temperature.

It will be unnecessary for me to enter into any
further detail on the nature of the vinous fermentation, as it is a chemical process, an account of
which, you may recollect, I have formerly given
you.

CONVERSATION XXX.

ON THE CULTIVATION OF GRASSES, TUBEROUS ROOTS, AND GRAIN.

———————

MRS. B.

THE objects of culture which we shall next investigate are the grasses.

The principal use of the grasses is to feed cattle, which, both during their life and after their death, are useful to us in so many different ways. The advantage we derive from them in agriculture is not confined to the labour they perform in the field; they also supply manure; and the more forage we produce for cattle, the greater is the quantity of manure we shall be able to spread upon our fields.

EMILY.

Poor soils, then, must require more cattle, and, consequently, more grass-land, than rich ones. But may not cattle be fed on other vegetables besides grass?

MRS. B.

Unquestionably; cattle will eat the same vegetables that serve for our subsistence; but we reserve these for our own use, and feed them on those which would afford us little or no nourishment, such as grasses. These are of two kinds, natural and artificial. The natural grasses are of the gramineous family, which belongs to the class of monocotyledons, or endogenous plants.

EMILY.

Yet how very little resemblance they bear to the palm-tree, or other tropical endogenous plants.

MRS. B.

They are not so dissimilar as you imagine, since they grow like them internally.

CAROLINE.

They may then be considered as the miniature palm-trees of our ungenial climates, being contracted both in space and time; for the mower comes with his destructive scythe, before they have passed through a single season.

MRS. B.

The natural grasses are either annuals or perennials. The first are very rarely used for meadows. In some countries, however, rye, Indian corn, and millet (all of which are annuals), are sown as grasses; that is to say, for the sake of their leaves, which are mown as soon as they appear above ground, and thus, several successive crops are obtained in one season.

But our meadows are all formed of perennial grasses : they are sown with hay-seed, which consists of a mixture of various sorts of grasses, more or less adulterated with the seed of weeds. These different grasses, ripening at different periods, a medium must be taken to mow the crop.

EMILY.

Would it not be better to sow only one species of grass ?

MRS. B.

Yes ; provided it were first ascertained, what species would best suit the soil and climate. There are some agriculturists, however, who dispute this opinion, and think that a variety of grasses makes the best fodder for cattle. Several naturalists are now engaged in endeavouring to raise very pure unmixed grasses, with a view to produce seed for sale: a measure which will greatly tend to the improvement of meadows.

CAROLINE.

Do not meadows occasionally require to be sown afresh ? for as the crops are either pastured or mown before the seeds are ripe, it cannot re-sow itself; and the grasses, though perennials, do not, I suppose, last a great number of years.

MRS. B.

Grasses are renovated, not so much by seed as by means of their roots and subterraneous branches, which spread out in various directions, interweaving and forming a sort of network of roots and branches; and from this reticulated mass spring abundance of new shoots, which thicken and renovate the meadow. If grass be kept short, it consumes less nourishment, and a greater quantity remains to push out fresh shoots.

EMILY.

This accounts for the fine thick turf of which our lawns are composed, and which, being so continually mown or fed off by sheep, precludes the possibility of their ever re-sowing themselves.

MRS. B.

This, however true in England, where the climate is temperate and moist, will not hold good in countries where the grass is burnt up in summer, when mowing cannot take place; and it is for this reason that it is impossible on the Continent to produce those beautiful lawns, so ornamental to our country seats.

These lawns, when first prepared, are not usually sown, but the grass is laid down in sods. By this means the roots are obtained ready matted, together with a thick fine turf, which it would require many years' growth, and constant mowing, to produce from seed.

Meadows are mown in England but once, or, at most, twice in the season; whilst, in many parts of the Continent, three or four crops are obtained, according as the soil is dry or moist, elevated or low.

The Phleum, the Dactylis, the Anthoxanthum, and Rye-grass, are the plants best adapted for meadows; but Rye-grass, degenerates in the dry warm climates of the Continent, as it requires a great deal of moisture to keep it fine and tender.

EMILY.

The great defect of grass, which I have observed both in France and Switzerland, is the quantity of weeds which are mixed with it, and which renders the hay strong and coarse.

MRS. B.

That is owing to the impurity of the seed, and is attended with every possible disadvantage. The coarse leaves of the weeds are not only unpalateable and unwholesome for cattle, but in growing they fill the spaces which the grasses would occupy, and, by separating them, prevent their roots from combining and giving rise to new shoots.

There are some meadows which, from peculiar circumstances, are not susceptible of being mown. The grass of mountains, for instance, does not grow sufficiently high to require it. Being frequently covered with mists, it remains green throughout the summer; much resembling our English lawns, and affording delicious pasture for cattle when the meadows in the valleys and plains are burnt up. The fine turf on the mountains of

Switzerland and the Alps, consists principally of
phleum, intermixed with other herbs of an inferior
quality. The matted roots of these grasses are
extremely useful in preventing the surface of the
soil from being washed down by rains: the meshes
of the network which they form confine the earth
and retain it, as it were, in a basket on the surface
of the declivity. It is on this account very im-
prudent to attempt tillage on the sides of mountains. Small flat patches of land, which are
occasionally met with in such districts, may be
cultivated with advantage, but it is dangerous to
displace the fence which Nature has provided; and
however inadequate the means may appear to the
end, it is certain that the massive mountains are
upheld, by the feeble roots of some of the smallest
of the vegetable species.

EMILY.

That is, indeed, wonderful! but it is merely the
surface of the soil which the roots of the grass
support.

MRS. B.

True; but if one surface were washed down,
another would be exposed to the same danger;
and thus, in the lapse of time, successive surfaces
would be destroyed, and the mountain finally be
brought low!

Another species of meadows incapable of being
mown are common fields, every parishioner having
a right of pasturage; a circumstance which renders

this species of tenure extremely disadvantageous: it is, in fact, condemning the land to yield as little produce as possible.

Let us now proceed to the artificial grasses, the most benevolent of all the vegetable tribe. It is to them that we are indebted for repairing the injury which the land sustains from the culture of grain. They were first introduced into France by the celebrated agriculturist, Olivier de Serres, in the sixteenth century.

EMILY.

These grasses do not, I suppose, form permanent meadows, but are sown alternately with crops of corn, in order to recruit the soil after the exhaustion it has undergone from the latter.

MRS. B.

Certainly: thus alternated, they form an excellent course of cropping. Most of the artificial grasses are of the leguminous family; among these the vetches and the scarlet clover are annuals. The common purple clover lasts two or three years. It is difficult to obtain the seed unadulterated by that of other plants. Clover has long been cultivated on the left bank of the Rhine, for the sole purpose of producing seed for sale: this commerce was chiefly carried on with England; for though we cultivate a considerable quantity of clover, we use it, almost wholly, as food for cattle; our summers seldom being hot enough

to ripen the seed, so that we are obliged to have
recourse to that of foreign growth. A very profit-
able trade was carried on with us in this article,
when Buonaparte issued his decree against export-
ation, and the poor agriculturists on the left bank
of the Rhine, then under the dominion of France,
were nearly ruined. The Germans, on the opposite
bank, supplanted them in a branch of commerce
they were compelled to abandon: England continued to be equally well provided with clover
seed; and it was Buonaparte's own subjects who
alone suffered by his absurd prohibition.

Saintfoin, or Esparcette, is another artificial grass
of the leguminous family, of longer duration
than clover. The seed appears larger than it
really is, because it is sown with the husk or pericarp, and no less than twenty pounds of seed is
required per acre; whilst ten or twelve pounds of
clover seed, which is sown without the husk, is
sufficient.

Lucerne, also of the leguminous family, lasts
from twenty to thirty years, according as the soil
is more or less favourable to it.

EMILY.

It is then too long lived to enter into a course
of cropping?

MRS. B.

In some parts of the valley of the Rhine, these
courses are made of thirty years' duration, twenty
of which is occupied by lucerne. The roots of

this plant strike twelve or eighteen feet into the soil; a depth at which moisture is always found, so that lucerne is enabled to resist drought much better than clover, whose roots are more superficial. Yet, if the season be dry, there is some danger of its failing the first year of its growth, the roots not having reached a depth of soil which is always moist. The seed for sowing should be chosen of a bright yellow colour, and heavy; a caution necessary to be attended to, in the choice of all seeds. Lucerne is mowed from three or four, to seven or eight times in the year, according to the climate in which it grows. Its herbage is less delicate than that of saintfoin.

CAROLINE.

Have we not seen lucerne growing as a shrub in some parts of Italy?

MRS. B.

This is of a different species; it is naturally a shrub, and grows wild on the sea-coast in Italy, where it is used as fodder for cattle. Furze may also be cultivated, either as a shrub or as artificial grass. In the latter state, it should be mown very young, while still soft and tender; after growing three years it is rooted up, but it prepares the soil admirably for grain.

There are some artificial grasses which are not leguminous; *Burnet* is of the family of

Rosaceæ : it has the advantage of thriving in calcareous soils.

The wild endive, and, indeed, the leaves of
almost any plant, are susceptible of being cultivated for forage ; excepting those which have
either milky or astringent juices, such as the leaves
of the fig, or the oak : cattle will not eat them ;
or at least not unless they are mixed with a considerable proportion of good forage.   But without
cultivating them as grasses, the young leaves of the
ash, the willow, and the acacia, gathered from the
tree, make very wholesome food for cattle.

EMILY.

In Italy, the cattle are very commonly fed on
the leaves of trees ; and I have often observed,
with admiration, the industry of the Tuscan peasants, who collect green weeds, the clippings of
hedges, and the leaves of trees, in order to supply
their cattle with food.

MRS. B.

The small size of the Tuscan farms, which
seldom exceed fourteen acres, do not admit of
meadow land, excepting the grass walks with
which they are intersected.   The nearer we approach the tropical climates, the more we find
meadows, both natural and artificial diminish : the
climate becomes too hot and dry for the cultivation of grasses.

We have observed, that there are some species of plants which afford food, both to men and cattle. These are the class of tuberous roots, which constitute one of the most valuable of the gifts of nature. The potato, the turnip, beet, and carrot, all belong to this class. Were these vegetables cultivated only in quantities sufficient to supply the wants of the human species, they would be considered as a most valuable acquisition, by varying, in a salutary and palatable manner, our stock of vegetable food. But when produced in such abundance that it is applied also to the sustenance of cattle, it not only extends the benefit to a lower order of beings, but furnishes, in case of need, a store of food for man ; the accidents which injure the crops of corn being seldom hurtful to tuberous roots.

CAROLINE.

And, should both fail, we have the resource of feeding on the cattle, who are themselves deprived of food.

EMILY.

It is this, no doubt, which explains what appeared to me very unaccountable, — that meat is sometimes cheap, when bread is dear.

MRS. B.

If forage fail, whether it be owing to a scarcity of grasses or of roots, a greater number of cattle will be sent to market, and the meat will consequently be low priced, but it will be of inferior quality; for, under such circumstances, the cattle cannot be fattened.

The culture of tuberous roots requires very deep ploughing. Beet is of various colours, most commonly of a rich crimson. It is raised from seed, and the young plants afterwards thinned: the soil should be neither very moist nor very dry: the seed ripens only the second year. This plant contains so great a quantity of saccharine matter, that, during the prohibitory system of Buonaparte, the French had recourse to it for the fabrication of sugar. Indeed, the manufacture is still carried on, and I understand that some recently-discovered mode of facilitating the process enables them to compete with the West Indian market.

There are three species of turnips; — *turnips*, *Swedish turnips*, and *the Kohl Rabi*, or turnip-rooted cabbage. The leaves of the first are rough and hairy, those of the second smooth, and those of the last form a medium between the other two, being hairy when young, and becoming smooth afterwards. There are many varieties of turnips; the white are the most delicate, the yellow the most robust: they require a light loose soil, and a good deal of manure; for being of the cruciform family, which contains *azote*, they must be furnished with the means of obtaining this element, and it is animal matter which yields it in greatest abundance.

The *Topinambour*, or Jerusalem artichoke, produces a great number of tubers, which are much
eaten in England; but they are not relished on the
Continent. This plant is cultivated in some parts
of America, and is brought to Europe from the
mountains of the Brazils.

Carrots require a light but not a loose soil:
they are rather of a delicate nature, suffering both
from excess of cold or of heat.

The potato, it is universally acknowledged, we
derive from America, but from what part is not
well ascertained; for it is remarkable that neither
M. Humboldt, nor any other traveller in that
country, has met with it in its wild state. Clusius,
the first botanist who speaks of potatoes, says that
they were introduced into Europe by the Spa-
niards, in 1588. Sir Walter Raleigh brought
them from Virginia to England and Ireland,
where their cultivation succeeded much better,
and they were more liked than on the Continent
of Europe; and it is we English, who have sub-
sequently been the means of introducing a taste
for them, into other countries.

EMILY.

In Italy the lower classes are still much pre-
judiced against potatoes, considering them as food
fit only for hogs or cattle.

MRS. B.

There are from one hundred to one hundred

and fifty varieties of this plant which differ in colour, form, precocity, &c. Potatoes are usually raised from germs, contained within the tuber, and commonly called eyes: these germs contain the rudiments of the young plant, similar to the buds or the branches of a tree. In order to make them sprout, the potato must be planted either entire or cut in pieces, leaving an eye in each piece, from which the young plant shoots; or in case of scarcity, the eye alone may be planted, reserving the fecula or mealy part for food.

EMILY.

I thought that the mealy part was a magazine of food for the young plants which shoot from these germs, and was, therefore, necessary to their developement.

MRS. B.

That is true; and the plant will shoot with much more vigour if the fecula remain attached to it; it is not, however, absolutely necessary: for the eye, if planted naked, has the power of absorbing moisture on which it feeds, till it has struck out roots, which supply it more regularly with nourishment. Potatoes may also be raised from slips, and as a last resource the seed may be sown; but this is so slow a process, that it is resorted to only with a view of procuring new varieties. Though the potato bears the name of tuberous root, the bulb does not grow upon the root of the plant but on the lower branches, which bury themselves

N 2

under ground: in cultivating the potato, it is necessary to hoe up the earth over these branches, in order to cover them more completely. There is a small tubercle produced by the potato-plant at the axilla of the leaf, which being exposed to the light becomes green, and is of so acrid a nature as not to be eatable. Half the weight of the potato consists in fecula: the saccharine principle may be developed in this tubercle as it is in barley; it will not produce wine, but spirit may be distilled from it when fermented; and the residue affords excellent food to fatten hogs. The other half of the potato consists in fibrin and mucilage. From eight to ten pounds weight of potatoes per day is a proper quantity to give to cattle when there is a sufficiency of hay to mix with it; but, in case of a scarcity of the latter, they will consume from eighteen to twenty pounds of potatoes without inconvenience.

We may now proceed to the examination of one of the most important of the vegetable productions in civilised countries,—I mean corn. We have hitherto considered gramineous plants as cultivated only for their leaves, under the name of grasses: but there are many of this family whose seeds are large enough to afford food for man, and it is with this view that he cultivates them. These are distinguished by the name of grain or corn, in Latin *Cerealia*, from Ceres the goddess of plenty, who is said to have first introduced corn into Sicily; but whence it originally came is unknown, it having never been found growing in a wild state. Some naturalists are of opinion that we derive grain from the mountains of Persia and Thibet; a species of wheat, the *Triticum Spelta*, commonly called *Spelt*, having been found growing wild in those countries. Others derive its origin from Tartary.

As corn belongs to the class of monocotyledons, the stems have no bark; but these tall and slender stalks derive their stability from a quantity of silex, which, not being of a volatile nature, is deposited on the surface of the straw or *culm*, when the more volatile parts evaporate. Here it accumulates, and in the course of time incloses the straw in a species of coat of mail, which not only enables it to resist injury, but also to support the weight of seed it has to bear.

EMILY.

Were it not for this provident supply of Nature, it is true that a slender hollow straw would be quite unequal to support the burden of a heavy ear of corn.

MRS. B.

In this and all northern countries, the straw is generally hollow, but in warm climates it is full. The stems of gramineous plants are also intersected with knots or articulations, designed, no doubt, to

N 3

add to its strength; and each of these, shoots out a long slender leaf, which encloses the stem like a sheath.

Grain constitutes the fruit of corn, and consists, consequently, of the seed and its pericarp: these are so closely attached together that they are not easily separated or distinguished from each other, when in the state of grain; but when ground into flour, it is the pericarp which forms the coarse bran; and the seed, the flour used for common household bread.

EMILY.

This flour then consists of the contents of the seed together with its spermoderm: and it is, no doubt, the latter which renders it brown?

MRS. B.

You are right; in order to obtain the whitest wheaten flour, such as the bread in London is made of, the spermoderm, which forms a finer species of bran, must also be subtracted: all this is very adroitly performed by that skilful naturalist the miller, with his sieve of moulting cloth.

EMILY.

How admirably this seed is protected! it is true that it is one of great importance to mankind: but is it not curious to think that so small a body as a grain of corn should have two coverings, consisting each of three coats?

CAROLINE.

And the husk, besides, for an outer garment. I thought it had been the husk which formed the bran.

MRS. B.

No, my dear; the husk constitutes the chaff which is separated from the grain by the operation of thrashing.

It is only in one species of corn, the *Triticum Spelta,* which I have just mentioned, that the husk adheres so firmly to the grain as to require a peculiar process of grinding, in order to part them. This renders it less liable to the depredations of the feathered tribe, who can easily pick out the naked grains of wheat from the ear; but find it very difficult to dislodge those of *epotre* from the adherent husk.

The seed contains the embryo plant and the albumen, which is to afford it the first nourishment, and this we have already said consists of fecula and gluten.

CAROLINE.

Since the albumen supplies so ample a provision for the young plant, the cotyledon of corn is not, I suppose, of a succulent nature?

MRS. B.

I beg your pardon; but it is so minute as to afford but very little sustenance.

The beard of corn is formed by the prolonga-

N 4

tion of the husks; it is not improbable that all species of grain was originally bearded, and that many of them lose this appendage when cultivated in good soil.

CAROLINE.

The beard, then, I dare say, is the result of a degenerated organ, like thorns or tendrils.

MRS. B.

Very likely; or at least that in a state of cultivation it disappears. Of the two species, bearded corn is by far the more robust; but it has the inconvenience of being subject to retain moisture, so that in a wet summer it is much more liable to injury.

Grain may be divided into three series:

First. That whose flowers have both pistils and stamens, and are aggregated in the form of ears.

Second. That with similar flowers, but in the form of clusters or bunches.

Third. That in which the pistils and stamens are situated in different flowers.

In the first series, which comprehends wheat, barley and rye, there are slits or cavities along the axis of the ear, whence issue smaller ears or earlets: in the spring these put forth a little flower and sometimes several, each of which contains a single grain, enclosed in a husk; these form the aggregated ear. The flowers have three stamens, and one pistil with two stigmas. The grain of wheat is of an oval form, that of epotres, rather triangular.

EMILY.

Is not the wheat sown in autumn more hardy than that which is sown in spring?

MRS. B.

There is no difference in them whatever, ex-cepting that the former ripens earlier.

The largest grains of corn should always be selected for sowing, because the pericarp does not increase in size in proportion to the seed.

It is the gluten contained in the grain of wheat which produces the fermentation of bread: this process is vulgarly called raising the bread; and it is true that the disengagement of carbonic acid, which takes place during fermentation, actually raises the dough, producing those hollow in-terstices which render bread light and digestible. Other species of corn do not make such good bread, as they contain less gluten.

EMILY.

I recollect, during a scarcity, potatoes being mixed with wheaten flour to make bread; but it rendered it very heavy and unpalatable.

CAROLINE.

I thought that yest, the produce of the ferment-ation of beer, was commonly used to excite that of bread.

MRS. B.

It is so in England; but it acts merely as a stimulus to hasten that of the gluten. On the Continent, and in wine countries in general, where beer is little drunk, the fermentation is excited by means of leaven, which consists of a piece of dough that has been kept from a former batch of baking, and has turned sour; or, chemically speaking, undergone the acetous fermentation. Now there is so much analogy between the acetous fermentation and that of bread, that it is sufficient to place a body which is undergoing, or has recently undergone the former, in contact with dough to excite it to ferment, and this may be done either with yest or leaven.

CAROLINE.

It is, then, a sort of contagion which these bodies communicate to the dough; but is it not surprising that it should render the bread light and wholesome, instead of turning it sour?

MRS. B.

Were the fermentation of the dough not interrupted by baking, it would become sour, as that portion does which is reserved for leaven. The fermentation of bread is by some chemists considered as a commencement of the acetous fermentation. There must, however, I conceive, be some difference between these processes, as in the regular succession of fermentations, the acetous is always subsequent to the vinous; and bread is so perfectly insipid that there is no reason to suppose it has undergone the latter.

EMILY.

Yet wheat is, I suppose, like other kinds of grain, susceptible of undergoing the vinous fermentation.

MRS. B.

Certainly; alcohol may be obtained from all kinds of grain. There are four species of wheat.

First. The common wheat, whose ears are erect, and its grains opaque and obtuse.

Second. The Triticum turgidum of Limoges, which the French call *Gros bled,* whose ears are thicker and larger; it contains less gluten, and, consequently, is not so well calculated for bread; but is much used on the Continent to thicken soup or porridge. This wheat, if cultivated in a very rich soil, produces a variety called *miraculous wheat,* the ears of which are branching from the abundance of their produce.

Third. *Bled dur,* or hard wheat: the grain is semi-transparent; it has still less gluten than the preceding: it is of this species that macaroni, vermicelli, and all the Italian pastes are made: it requires a dry soil and a warm climate, and thrives best in the southern parts of Europe.

Fourth. Polish wheat: it grows very plentifully in Poland, and is thence exported to other countries; but being of inferior quality, it is little cultivated elsewhere.

N 6

Spelt contains less gluten than other species of wheat: it affords beautifully white flour for pastry, and is also much used for starch.

EMILY.

I should have thought that it would have required more gluten to make starch than to make bread?

MRS. B.

No; starch consists almost wholly of pure fecula, and may be obtained from potatoes as well as from wheaten flour.

Rye is of so hardy a nature that it accommodates itself to almost all soils and all climates: its straw is longer and firmer than that of wheat, which renders it peculiarly adapted to thatching: it contains so little gluten that it cannot be made into bread without an admixture of wheat.

EMILY.

It is, then, no doubt, on this account that the poor Scotch Highlanders, who cannot afford to mix wheaten flour with it, eat it baked in cakes

MRS. B.

It is chiefly oats, I believe, that are thus eaten in Scotland.

Barley is principally used for fermentation. It contains a great quantity of saccharine matter: mixed with hops we have seen that it produces beer; and it is also distilled for spirits.

13

In the second series of corn, the grain grows in the form of clusters, each earlet having a separate pedicle or footstalk.

This series contains four genera.

First. Oats: the husks or glumes have two valves and beards springing from the back part of the husk, instead of growing from the summit, as with barley and rye. Oats afford food both for man and for horses.

Second. A species of oats derived from Asia, the earlets of which incline all in one direction: it is more robust than the preceding, yet it is very liable to be attacked by the disease called *smut*.

Third. The Phalaris of the Canary Isles, commonly called Canary seed, used chiefly as food for the birds, for which those islands are so celebrated.

Fourth genus. Rice, which we derive both from the East and West Indies. Next to the Banana or bread tree, this is the plant which affords the greatest quantity of wholesome nourishment, and is, perhaps, susceptible of the greatest variety in the mode of cooking; for being itself insipid, it admits of all kinds of seasoning.

EMILY.

How much, then, it is to be lamented that its cultivation should be unwholesome!

MRS. B.

It is so only at that period when the water is

drawn off to enable the grain to ripen. It is sown in the spring in a muddy soil; and as the plant grows, the water is let on, and gradually raised so as to keep it almost wholly covered, until the grain begins to ripen. I have been informed that in the rice plantations of Lombardy, the mortality is not greater than in the adjacent districts: it is true that the inhabitants of the latter are in a wretched state of poverty, whilst the cultivators of rice are at least supplied with plentiful nourishment, to compensate for the unwholesomeness of their occupation. I should not wish to extend the culture of rice in Europe, in soils adapted to other produce, but, as this plant will grow only in marshy districts, it is as well to convert such land to so useful a purpose; for it is not more unhealthy as rice fields, than as marshes. One great objection to the cultivation of rice is, that it injures the surrounding soil by the filtration of the waters, which, in the course of time, destroys all the trees in the neighbourhood.

CAROLINE.

But such a filtration must be very advantageous to meadow land?

MRS. B.

When confined within due limits; but we must remember the old adage, " Too much of a good thing is good for nothing:" the adjoining meadows would eventually become converted into marshes, so that there would be no other resource than to

14

extend the cultivation of rice; and the evil would thus always go on increasing, if government did not interfere to prevent it.

The third series of corn, having the pistils and stamens in different flowers, consists of maize or Indian corn, Canada rice, Sorghum, or millet.

We have run through a great variety of subjects to-day; the natural and artificial grasses, which form the two great stores of food for cattle, and the latter of which enters so beneficially into the rotation of cropping; the tuberous roots of which both man and beast equally partake; and, finally, the numerous species of grain, which afford a more solid and wholesome nourishment than any other kind of vegetable food. After such an abundant store of sustenance, were I to prolong the subject further I fear it would satiate your appetite, we will therefore reserve what remains to be said on vegetable food till our next interview.

CONVERSATION XXXI.

ON OLEAGINOUS PLANTS, AND CULINARY VEGETABLES.

———————

MRS. B.

AMONG the articles of vegetable food, the oils which are extracted from plants afford one of the most valuable; nor are they of less importance in affording us light by their combustion. They are employed also in a number of manufactures, such as soap, woollens, varnishes, and perfumery.

There are, you know, two kinds of vegetable oil, distinguished by the name of fixed and volatile. The latter may be extracted from almost every plant; but it is used only as a perfume, or to flavour liqueurs, such as the oil called attar of roses, with which you are acquainted.

EMILY.

Yes; and with that of jasmine and orange, so commonly used to perfume pomatum. In a word, the perfumer's shop abounds with these sweet-scented oils.

MRS. B.

They constitute the luxury of the sense of smelling, but are frequently prejudicial from their effect on the nerves ; and some few of them are employed medicinally.   But the essential or volatile oils are not those most deserving our attention : the fixed oils are of much higher importance, and are extracted from a class of plants, hence called oleaginous.   The oil is expressed from the seed of all these plants excepting the olive, in which it is obtained from the pericarp, or fleshy part of the fruit which surrounds the seed.

The greater part of the seeds of oleaginous plants contain albumen, and it is from this, that the oil is obtained ; but when the seed has no albumen, as is the case with the poppy, it is the embryro which furnishes the oil.

In the family of the *Euphorbiaceæ,* all of which have oleaginous seeds, the embryo is of a venomous natúre, and oil extracted from it would be poisonous ; while that expressed from the albumen of the same plant, situated contiguous to the embryo, is perfectly innocent.   Such is Bancul nut (*Aleurites Moluccanum*), which is remarkably mild, and is eaten by the inhabitants of the Molucca isles, as we eat hedge-nuts in Europe, while oil obtained from the embryo is an acrid poison.

EMILY.

Can oil be expressed from plants growing wild,

or is it necessary they should be cultivated in order
to supply it?

MRS. B.

Some small quantity may be obtained from
thistles : the stone pine, and plum-tree of Brian-
con also yield it; but it is the seed of the beech-tree
alone which affords it in sufficient abundance to
make it worth the labour of obtaining. The forest of
Villers-Coterot, in France, produces a great quantity of this oil. It is less liable to become rancid
than any other, and, on this account, is often
mixed with olive oil, which is to be exported to
America or any other distant part; but it all passes
under the name of olive oil.

The fixed oils obtained by cultivation may be
ranged under three heads: first, olive oil, the produce of warm climates; secondly, nut oil, that of
temperate climates; and, thirdly, oils obtained from
the seeds of oleaginous herbs.

The olive-tree originally came from Syria. That
plant, as well as the vine, was brought to Marseilles
by the Phocians; and, at the present day, it is cultivated on all the shores of the Mediterranean. It
is a tree of very slow growth, but of long duration:
it can support a temperature as low as eight or ten
degrees of Fahrenheit, provided the air be dry; but,
if accompanied with humidity, one or two degrees
below the freezing point, proves fatal. The plant,
however, may recover, if cut down to the roots, a
little below the surface of the soil; it then strikes
out fresh shoots, forming five or six young trees.

Manure used for olive-plantations should be of a dry nature; and it is necessary to heap up the earth over the roots, to keep them well covered.

CAROLINE.

These roots must be naturally very superficial; for, notwithstanding the care that is taken to cover them with earth, I have observed that they are continually making their appearance above ground.

MRS. B.

It is rather the rugged and tortuous base of the stems which you have observed, and which wear the appearance of roots.

There are several varieties of olive-trees. Those of the plantations about Nice, afford us oil perfectly white and limpid, and equally free from either smell or taste : it is held in very high estimation in northern countries, where the natural taste of oil is disliked, probably from its being associated with that of rancidity; but, in the countries which produce oil, where, being eaten fresh, it is very seldom rancid, the oil which partakes of the flavour of the fruit is preferred.

The fruit should be gathered, not shaken from the tree in order to prevent their being bruised, and the oil expressed as soon as possible afterwards, otherwise there is danger of rancidity. In Spain, and other countries where feudal tenures still exist, the olive-mills belong to the lords of the land, and the peasantry are obliged to wait their turn for their olives to be pressed, to the great detriment of the produce. This is, perhaps, the only harvest which is gathered in about Christmas, the fruit not being ripe earlier.

Olives begin to be cultivated at 43° of latitude: in tropical climates, they will grow at two hundred fathom above the level of the sea.

EMILY.

And, in temperate climates, where the olive ceases to grow, the walnut replaces it.

MRS. B.

Yes; but the oil obtained from the walnut is far inferior to that of the olive, having both colour, smell, and flavour, qualities which are not esteemed in oils. The walnut-tree succeeds better in a northern than southern aspect; for, as the young shoots are very liable to suffer from a white frost, it is desirable that their vegetations hould be retarded till the spring is so far advanced, that there will be little danger of their encountering that evil. This tree grows remarkably well at the foot of a mountain, on account of the depth of soil produced by the quantity of earth washed down.

The cultivation of oleaginous herbs enters into the course of cropping: they exhaust the soil almost as much as grain, on account of the number of seeds to be ripened; they require, therefore, a considerable quantity of manure. These herbs are generally of the cruciform family, containing azote, an element of the animal kingdom which forms excellent manure : so that, after the oil is expressed, the cake which remains serves to restore the exhausted soil. Rape is a species of cabbage with thin roots, whose seeds yield excellent oil.

The poppy is an oleaginous plant, with white, scarlet, and violet flowers, while the seeds are white or black. They yield oil, perfectly innoxious and wholesome, though drawn from the same plant which supplies us with opium.

CAROLINE.

I confess I should always be apprehensive of its being adulterated with some mixture of its poisonous neighbour. Is not flax, also, an oleaginous herb ?

MRS. B.

Yes. It is, however, chiefly cultivated for its stalks, from which linen thread is fabricated; but its seed also yields the oil we call linseed-oil. It is much used in the art of painting. Hemp is of the same description. There are some few oleaginous herbs of the leguminous family, such as the subterranean Arachis (*Arachis hypogæa*), a plant we derive from America, which has the singular property of ripening its seeds under ground. This plant requires a loose sandy soil, in order that the lower branches may be enabled to bury themselves in the ground. In a state of cultivation, the earth should be heaped over them, as is done with potatoes. The upper branches, which

blossom in the air, ripen no seed ; while the lower
lateral branches, which burrow in the earth, develope no regular blossom, that is to say, have no
petals; but the stamens and pistils bring the seeds
to perfection.

Among the objects of cultivation, the vegetables
raised in our gardens for culinary purposes form
a class of considerable interest.

CAROLINE.

In our choice of these, we must be regulated
by the palate.

MRS. B.

Principally, no doubt; but modified by other
circumstances, such as soil, climate, &c. Plants
of a fibrous, woody nature are too tough to be
either palatable or digestible. Those which are
acrid, or very bitter, must equally be rejected. A
powerful flavour is also objectionable; and, on the
other hand, great insipidity will not gratify the
palate. Here, then, are many causes of exclusion,
but some of them admit of remedy.

Plants of an acrid nature may be eaten young,
before the acridity is well developed, especially if
the most delicate parts be chosen, which are those
that have been least exposed to the light. Thus
the receptacle, or what is commonly called the
bottom of the artichoke, and the internal part of
the bractæ, are mild and pleasant to the taste when
young.

EMILY.

And asparagus we eat as soon as the young shoots appear above ground.

CAROLINE.

And do we not even find the taste of rhubarb in tarts delicate and pleasant, when the plant is very young, while, when full grown, it is so repugnant to us?

MRS. B.

It is true that rhubarb requires to be eaten very young, in order to be palatable; but it is the ribs of the leaves which we make into tarts, while that part of the plant taken medicinally, and which is so pungent and disagreeable, is the roots; and it must grow in a warmer climate to have its medicinal properties developed. That which we import from Turkey is grown either in Tartary or at the foot of Mount Caucasus.

All strong vegetable flavours, even that of the Prussic acid, which is one of our most deadly poisons, may be rendered agreeable to the palate, and perfectly innocent, if taken in very minute portions, and mixed up with considerable quantities of insipid food. The Prussic acid is found in the kernels of peach-stones and in bitter almonds, but in very small quantities; and yet one or two of these is sufficient to communicate an exquisite flavour to a dish of cream or of pudding.

Celery belongs to the class of Conium, or hemlock, the poison which caused the death of Socrates; but its pernicious qualities will not be developed, and it will grow white and tender, if the stems be kept covered with earth.

EMILY.

Insipid plants should, then, on the contrary, be fully exposed to the light and air, in order to bring forth what little flavour they contain?

MRS. B.

Yes; and they should be eaten only when full grown. Great insipidity is not wholesome, any more than a very strong flavour: the one produces too great excitement in the digestive organs, the other does not afford them sufficient stimulus.

EMILY.

Both these defects, I should think, might be corrected, by cooking vegetables of such opposite qualities together.

MRS. B.

It is with this view that thyme, sage, mustard, onion, and even garlic, are used as seasoning for food of an insipid nature; and sugar and spices are most useful auxiliaries for such a purpose.

CAROLINE.

Salt seems to be the most universal of all ingredients to season cookery.

I omitted mentioning it, because it was not of the vegetable kingdom.

There are no less than fifty-four species of plants, which may be considered as belonging to the class of culinary vegetables. These are derived from thirty-nine genera and seventeen families; and produce above five hundred varieties.

Among these families, the Cruciform supplies our table with the greatest number of dishes. It derives its name from the blossom having four petals in the form of a cross. Azote is found in this family alone, and it communicates to the vegetables a strong flavour, and often an offensive smell. The various species of cabbages belong to it, such as the common cabbage, the curled cabbage, broccoli, cauliflowers, turnips, radishes, water-cresses, and sea-kale.

Do you include turnips and radishes among the species of cabbages?

Their leaves and blossoms are of the same description; but the appearance of the vegetables on table, I confess, is totally different; and no wonder, for in the one it is the leaves we eat, in the two others the roots.

CAROLINE.

The leaves of the turnip, it is true, would be too strong and pungent for our palate. They are relished by sheep and cattle; and the root, which is more delicate from not being exposed to the light, is better suited to our taste.

EMILY.

The roots of radishes are, however, so strongly flavoured, as to be disagreeable, unless eaten very young. In the cauliflower it is the blossom, and not the leaves, that we eat.

MRS. B.

The head of a cauliflower has, it is true, much the appearance of a blossom, but it consists only of numerous ramifications of the peduncles, or flower-stalks, which not having sufficient space to grow in, adhere together, and form the white mass which we esteem as a very favourite dish of vegetable food.

EMILY.

But the cauliflower is rather of an insipid than of a pungent nature, and requires salt to season it.

MRS. B.

Its flavour is not strong if the head only be eaten; but the smell and taste of the water in which it is boiled is extremely offensive, and that of the vegetable itself is often unpleasant, when served at table.

EMILY.

I know scarcely any odour more disagreeable than that proceeding from a plantation of decayed cabbages, in which the azote is fully developed.

MRS. B.

When the cauliflower is allowed to attain its natural growth, or, as the gardeners express it, is left to run to seed, the flower-stalks lengthen and spread, and the blossoms are developed at their extremities. Broccoli is of a similar nature : the pedunculi amalgamate and form a head ; but it is of a green colour, because not so closely enveloped in leaves and sheltered from the light as the cauliflower. The small tender grain which is deposited upon it, consists of the embryo of blossoms which cannot be developed, owing to the quantity of nourishment of which the stalks deprive them.

The Leguminous family affords us four species of culinary vegetable, — peas, beans, lentiles, and kidney-beans; of some of these we eat only the seeds; in others, such as the kidney-bean and sugar-pea, the pod or pericarp are also eaten.

The family of Cucurbitaceæ supplies us with cucumbers, pumpkins, and melons : the two first are rather arbitrarily denied the name of fruit, and are ranked as culinary vegetables, merely on account of the saccharine principle not being developed in them.

This family is distinguished by a bitter principle contained in one of its species, the Colocinth: it is so strong as to be taken only medicinally.

From the Umbelliferous family we obtain carrots, parsley, lettuces, and hemlock. The narcotic principle exists throughout this family: in hemlock it is so powerful as to constitute a poison; but in most of the other species, it exists in such small quantities as not to be deleterious.

The family of Solanum gives us the Potato, Tomata, and the Belladonna, celebrated for the poison it contains.

CAROLINE.

And yet nearly akin to the potato, which is of so innocent a nature!

MRS. B.

That is true of the tubercle we eat, but the fruit of the plant is of an acrid nature: you may probably have been warned, in your childhood, of the poisonous properties of the small green tubercles which grow on the branches.

The family of Fungi supplies us with the mushroom, a vegetable of a most delicate and exquisite flavour; but as those species which grow wild are generally of a poisonous quality, it is important that we should learn how to produce such as are known to be innoxious. For this purpose, the white filaments, commonly called the spawn of mushrooms, should be cut in pieces and sown in a hot-bed. Whether these filaments consist of shoots, runners, or seeds of mushrooms, has not been well ascertained ; but when spread over a hot-bed, and sheltered from the open air, either under a shed or in a cellar, they will germinate. In Paris, mushrooms are raised in the Catacombs; and I know no place where they are produced in such abundance, or sold so cheap. The spawn should be sown in December, covered with a little earth and a litter of straw, then watered; and after a short time, if the litter be raised, the mushrooms will be seen growing beneath it.

These are some of the principal families from which we derive our vegetable food: I will not attempt to go through the whole seventeen,—it would be uselessly trespassing upon your patience.

I have now, I believe, imparted to you the whole of my little stock of botanical knowledge. The source from which I drew it was rich and copious ; but I am too well aware of my incapacity to do justice to the subject, not to shrink at the apprehension of having disfigured those lessons which afforded me such a delightful source of instruction; which taught me to investigate, with wonder and admiration, the beautiful organisation of the vegetable creation, and raised my mind, with increased fervour of gratitude, towards their bountiful Author.

INDEX.

A

o 4

Branches, ii. 4. 14. 51.
Brandy, ii. 241.
Bracteæ, i. 73.
Bread, ii. 273.
Bread-tree, ii. 81.
Brocoli, ii. 289. 291.
Broom, i. 229. 231. 279.
Bryophyllum, ii. 63.
Buck-wheat, i. 271.
Buds, i. 76. ii. 26. 175.
Bulbous root, i. 31. 44. 79.
Burnet, ii. 262.

C

Cabbages, i. 145. ii. 289.
Cacti, ii. 160. 179. 198.
Calendar of Flora, ii. 40.
Calyx, ii. 30. 52. 55. 73.
Cambium, i. 21. 112. ii. 7. 193.
Campine, i. 230.
Canadyria, ii. 279.
Caoutchouc, i. 127.
Capsular fruits, ii. 82.
Carbon, i. 102. 120. 174. ii. 190. 221.
Carbonic acid, i. 138. 174. 244. 248.
Cardoons, i. 145.
Carnations, i. 82.
Carpel, ii. 34. 72.
Carrot, ii. 264.
Castanea vesca, ii. 79.
Cattle, i. 270.
Cauliflower, ii. 289. 290.
Celery, i. 145. ii. 287.
Cellular system, i. 14. 48. 54. ii. 149.
Cereala, ii. 268.
Chaff, ii. 51.
Charcoal, i. 251.
Cherry, ii. 237.
Chesnut, i. 53. ii. 79.
China aster, i. 80. ii. 47.
Cider, ii. 241.
Cinchona, ii. 160.

Classification of plants, i. 135. ii. 107. 132.
———— artificial systems of, ii. 119.
———— natural ditto, ii. 135.
———— Linnæus's, ii. 126.
Climate, i. 160.
Clover, i. 274. ii. 260.
Cochineal, ii. 198.
Cocoa-nut, i. 42. 45. 79. ii. 63.
Colmare, i. 216.
Colocinth, ii. 292.
Coloured leaves, i. 73.
Colours of plants, i. 138. ii. 171.
Columella, ii. 72.
Coma, ii. 55.
Combler, i. 216.
Compound leaves, i. 68.
Compound flowers, ii. 47.
Cone, ii. 78.
Conium, ii. 287.
Contractibility, i. 10.
Copal, i. 128.
Core, ii. 75.
Cork, i. 56.
Corn, i. 44. 270. ii. 2076
——, beard of, ii. 271.
——, series of, ii. 272.
——, ears of, ii. 272.
——, flowers of, ii. 272.
Corolla, ii. 31. 55.
Cortical, i. 47.
Cotyledons, i. 41. 71.
Couch-grass, i. 28.
Cow-tree, i. 127.
Cratægus crusyalli, ii. 235.
Creeping plants, ii. 200.
Cruciform, ii. 289.
Cryptogamous plants, i. 181.
Cucumbers, ii. 291.
Cucurbitaceæ family, ii. 291.
Culinary vegetables, ii. 285.
Culm, ii. 269.
Cuscuta, ii. 205.
Cuticle, i. 18. 56. ii. 84.
Cyme, ii. 39.
Cyneps, ii. 43. 197.

Hybrid, ii. 168.
Hydrangea, i. 73. 122. 140.
Hydraulic rain, i. 206.
Hygrometric power, i. 14.

I & J

Jerusalem artichoke, ii. 266.
Jessamine, i. 48.
Ilex, ii. 235.
Indian fig-tree, i. 34.
Inflorescence, ii. 40.
Ink, ii. 197.
Insects, ii. 195. 197.
Inundations, i. 218.
Involucrum, ii. 38. 49.
Irrigation, i. 202.
Irritability, i. 10.
Ivy, ii. 200.

K

Kali or kelpwort, i. 179. ii. 165.
Kalmia, i. 166.
Knotted root, i. 82.

L

Labiate flowers, ii. 59.
Lakes, i. 188.
Laurientius, ii. 205.
Layers, ii. 2. 6.
——, season for, ii. 11.
Leaf-buds, i. 77.
Leaven, ii. 274.
Leaves, i. 64. ii. 61. 64.
——, sessile, i. 66.
——, articulated, i. 66.
——, fibres of, i. 66.
——, ribs of, i. 66. ii. 62.
——, stoma of, i. 67.
——, pennated, i. 67.
——, palmated, i. 67.
——, peltate, i. 67.
——, pedatum, i. 67.

Leaves, simple ribs, i. 67.
——, contour of, i. 68.
——, pinnatifid, i. 68.
——, dissected, i. 68.
——, compound, i. 68.
——, succulent, i. 69.
——, seminal, i. 71.
——, primordial, i. 72.
——, radical, i. 73.
——, floral, i. 73.
——, coloured, i. 73.
——, arrangement of, i. 75.
——, folding of, i. 85.
——, deciduous, i. 86.
——, fall of, i. 87. 123. ii. 191.
——, variegated, ii. 186.
Leguminous crops, i. 270.
—— family, ii. 289.
Lentiscus, ii. 235.
Lettuces, i. 144. ii. 292.
Liber, i. 116.
Lichens, i. 17. 42. 181. ii. 199.
Light, i. 95. 137.
Ligneous, i. 47.
Ligulate florets, ii. 57.
Liliaceous plants, i. 44.
Lime, i. 90. 122. 243.
Linseed oil, i. 180.
Lobes of the seed, i. 70.
Lopping trees, ii. 227.
Lucerne, ii. 205. 261.

M

Madder, i. 229.
Magnesia, i. 90. 122.
Magnolia, ii. 10. 79.
Maize, i. 208. ii. 279.
Malaxis paludosa, ii. 63.
Malvaceous, ii. 147.
Mangrove, ii. 9.
Manna, i. 126. 129. 190.
Manure, i. 232. 237. 247.
——, short, i. 253.
——, long, i. 253.
Marl, i. 243.

302   INDEX.

Rosaceæ, ii. 237. 263.
Rose-tree, ii. 206.
Rot, ii. 207.
Rotation of crops, i. 257. 274.
Rushes, i. 82.
Rust, ii. 208.
Rye, i. 282. ii. 208. 276.
—— grass, ii. 258.

S

Saffron, ii. 206.
Saintfoin, ii. 261.
Salicaria, ii. 165.
Salicornia or saltwort, ii. 165.
Salsify, ii. 49.
Salsola, i. 179. ii. 165.
Salts, i. 122.
Sand-hills, i. 228. 231.
Sands, i. 218.
Sap, i. 15. 60. 89. ii. 192. 240.
—— rise of, i. 61.
—— composition of, i. 90. 101.
—— exhalation of, i. 94.
—— descent of, i. 112.
—— velocity of, i. 63. 90.
Sarmentaceæ, ii. 242.
Saxote, ii. 165.
Scales, i. 76.
Scarzonera, ii. 58.
Scions, ii. 6. 12. 17. 26.
Sea-cale, ii. 289.
Sea-salt, i. 181.
Secretions, internal, i. 126.
———— excretory, i. 130.
Seed, i. 20. 60. 181. 193. 224.
    266. ii. 2. 29. 32. 54. 80. 83.
    178. 210. 248. 271.
—— structure of, ii. 83.
——, conveyance of, ii. 157.
Seminal leaves, i. 71.
Sensibility, i. 9.
Sensitive plant, i. 9.
Sepals, ii. 30.
Shoot, i. 60.

Silex, i. 90. 122.
Siliques, ii. 71.
Sleep of plants, i. 137.
Slip, ii. 2. 6. 12.
Smoke, ii. 190.
Smut, ii. 207.
Sobanum family, ii. 292.
Social plants, ii. 164.
Soda, i. 90. 122.
Soil, i. 222.
—— argillaceous, i. 226.
—— silicious, i. 227.
—— improvement of, i. 232.
    240.
Soldering, ii. 145.
Soot, i. 251.
Storghum, ii. 279.
Spade, i. 233.
Spanish chesnut, ii. 79.
Species, ii. 109. 150. 167.
Spelt, ii. 269.
Spermoderm, ii. 55. 83. 270.
Spike, ii. 39.
Spindle-shaped root, i. 29.
Spongiole, i. 22. 25.
Stamens, ii. 35. 45. 55. 114.
    179.
Standards, ii. 218.
Stapedra, i. 108.
Starch, ii. 276.
Station of plants, ii.153.
Stem, i. 37.
—— subterraneous, i. 39.
—— endogenous, i. 41.
—— exogenous, i. 41.
—— structure of, i. 42.
—— germs of, ii. 4.
Steppes, i. 128.
Stigma, ii. 34.
Stipula, i. 68. 78.
Stock, ii. 17.
Stoma, i. 67. 94.
Stones, i. 240.
Stone-fruit, ii. 67.
Straw, ii. 269.
Strawberry, ii. 76.
Style, ii. 34.

THE END.

LONDON:
Printed by A. & R. Spottiswoode,
New-Street-Square.

Printed in the United States
By Bookmasters